"This exciting collection of original articles addresses both theoretical and practical questions about toleration. It does so in ways that throw fresh light on the complex problems of understanding and negotiating the 'spaces', both literally and metaphorically, for the practice of tolerance in a wide variety of contexts, with particular but not exclusive reference to the sphere of religious toleration. Anyone who is seriously interested in toleration should find plenty to think about in this stimulating and sometimes provocative book".

– John Horton, *Professor Emeritus of Political Philosophy, Keele University, UK*

Spaces of Tolerance

This book offers interdisciplinary and cross-national perspectives on the challenges of negotiating the contours of religious tolerance in Europe.

In today's Europe, religions and religious individuals are increasingly framed as both an internal and external security threat. This is evident in controls over the activities of foreign preachers but also, more broadly, in EU states' management of migration flows, marked by questions regarding the religious background of migrating non-European Others. This book addresses such shifts directly by examining how understandings of religious freedom touch down in actual contexts, places, and practices across Europe, offering multidisciplinary insights from leading thinkers from political theory, political philosophy, anthropology, and geography. The volume thus aims to ground ideal liberal democratic theory and, at the same time, to bring normative reflection to grounded, ethnographic analyses of religious practices. Such 'grounded' understandings matter, for they speak to how religions and religious difference are encountered in specific places. They especially matter in a European context where religion and religious difference are increasingly not just securitised but made the object of violent attacks.

The book will be of interest to students and scholars of politics, philosophy, geography, religious studies, and the sociology and anthropology of religion.

Luiza Bialasiewicz is a political geographer and Professor of European Governance at the University of Amsterdam, where she also co-directs the Amsterdam Centre for European Studies (ACES). Her research focuses on European migration policy, within European cities and at the borders of Europe.

Valentina Gentile is Assistant Professor in Political Philosophy at LUISS University of Rome and Guest Professor at Antwerp University. She specialises in normative political philosophy, liberal theory, and the work of John Rawls. Her research focuses on stability, pluralism and reciprocity, tolerance, and equality.

Routledge Research in Place, Space and Politics
Series Editor: Professor Clive Barnett
University of Exeter, UK

This series offers a forum for original and innovative research that explores the changing geographies of political life. The series engages with a series of key debates about innovative political forms and addresses key concepts of political analysis such as scale, territory and public space. It brings into focus emerging interdisciplinary conversations about the spaces through which power is exercised, legitimized and contested. Titles within the series range from empirical investigations to theoretical engagements and authors comprise of scholars working in overlapping fields including political geography, political theory, development studies, political sociology, international relations, and urban politics.

Direction and Socio-spatial Theory
A Political Economy of Oriented Practice
Matthew G. Hannah

Postsecular Geographies
Re-envisioning Politics, Subjectivity and Ethics
Paul Cloke, Christopher Baker, Callum Sutherland and Andrew Williams

Oil, Culture and the Petrostate
How Territory, Bureaucratic Power and Culture Coalesce in the Venezuelan Petrostate
Penélope Plaza Azuaje

Migration in Performance
Crossing the Colonial Present
Caleb Johnston and Geraldine Pratt

Spaces of Tolerance
Changing Geographies and Philosophies of Religion in Today's Europe
Edited by Luiza Bialasiewicz and Valentina Gentile

For more information about this series, please visit: www.routledge.com/series/PSP

Spaces of Tolerance

Changing Geographies and Philosophies
of Religion in Today's Europe

**Edited by Luiza Bialasiewicz and
Valentina Gentile**

LONDON AND NEW YORK

First published 2020
by Routledge
2 Park Square, Milton Park, Abingdon, Oxon OX14 4RN

and by Routledge
52 Vanderbilt Avenue, New York, NY 10017

Routledge is an imprint of the Taylor & Francis Group, an informa business

First issued in paperback 2021

British Library Cataloguing-in-Publication Data
A catalogue record for this book is available from the British Library

Library of Congress Cataloging-in-Publication Data
A catalog record for this book has been requested

ISBN: 978-0-367-22407-3 (hbk)
ISBN: 978-1-03-208765-8 (pbk)
ISBN: 978-0-429-27473-2 (ebk)

Typeset in Times New Roman
by Apex CoVantage, LLC

Contents

Figures

Contributors

Peter Balint is Senior Lecturer in International & Political Studies at UNSW Canberra, and Convenor of the International Ethics Research Group. He is a political theorist, whose research centres on the issues of diversity (including toleration, respect, neutrality, and national identity), as well as on issues of privacy in the information age. His books include *Respecting Toleration: Traditional Liberalism & Contemporary Diversity* (Oxford University Press, 2017), and *Liberal Multiculturalism and the Fair Terms of Integration* (edited with Sophie Guérard de Latour, Palgrave, 2013). He is a founding member of The Global Justice Network and founding Editor of its journal, *Global Justice: Theory Practice Rhetoric.*

Luce Beeckmans is a post-doctoral research fellow funded by the Flanders Research Foundation (FWO) and affiliated to Ghent University (Department of Architecture and Urban Planning) as well as KULeuven University (Interculturalism, Migration and Minorities Research Centre) and Antwerp University (Urban Studies Centre) in Belgium. In 2005, she graduated as an engineer-architect at Ghent University and then worked at the international office of Stéphane Beel Architects. In 2013, she obtained an award-winning dissertation from the University of Groningen (Groningen Research Institute for the Study of Culture, ICOG) in which she studied colonial and post-colonial urban development in sub-Saharan Africa from a comparative and interdisciplinary perspective. Luce Beeckmans has a keen interest in the circulation of spatial knowledge between Europe and Africa, both within the framework of colonisation and development cooperation and as a result of trans-national migration. Her most important research topics are urban segregation and encounter; urban agency and citizenship; spatial appropriation and place-making; housing and diversity; and global religious spatiality. She has published widely and co-curated exhibitions on these topics.

Luiza Bialasiewicz is a political geographer and Professor of European Governance at the University of Amsterdam, where she also co-directs the Amsterdam Centre for European Studies (ACES). She has been a visiting professor at the College of Europe (Natolin) since 2013, where she teaches an annual course on European geopolitics. Her work focuses on the political geographies of European foreign policy and EU geopolitics, with particular attention to the EU's external borders and the out-sourcing of migration management. Her most recent research has been focussed on the changing spatialities of migrant reception within European cities and on far-right movements in a number of European states. Her previous edited book is *Europe in the World: EU Geopolitics and the Making of European Space* (Ashgate, Critical Geopolitics Series).

Margherita Galassini is a PhD candidate in Political Theory at LUISS Guido Carli University in Rome and a junior visiting scholar at Nuffield College, University of Oxford. Her research focusses on questions of religious accommodation, state legitimacy and the role of the Church in Catholic countries.

Valentina Gentile is Assistant Professor in Political Philosophy at LUISS University. From 2016 she has been Guest Professor at Antwerp University, holding a chair in "Interreligious Dialogue and the Ethics of Peace". Her research focuses on normative political theory, moral stability, pluralism, and the principles of reciprocity, toleration, equality, and social justice. Her recent publications include (2018) "From a Culture of Civility to Deliberative Reconciliation in Deeply Divided Societies" in the *Journal of Social Philosophy*; (2018) "Rawls's inclusivism and the case of 'religious militants for peace': A reply to Weithman's restrictive inclusivism", in *Philosophy and Public Issues (New Series)*; (2017) "Democratic Justice: The Priority of Politics and the Ideal of Citizenship" in *CRISPP Critical Review of International Social and Political Philosophy*. She is the co-editor of the collection *Rawls and Religion* (Columbia University Press), and author of *From Identity-Conflicts to Civil Society: Restoring Human Dignity and Pluralism in Deeply Divided Societies* (LUISS University Press).

William Haynes is an urban geographer and Junior Research Fellow in the Amsterdam School for Regional, Transnational and European Studies. His work explores the informal spatialities of migrant reception as well as contestations over public space in European cities by migrant groups, with a particular focus on train stations. He will be starting his PhD in Geography at the University of Sheffield in Autumn 2019.

Peter Jones is Emeritus Professor of Political Philosophy at Newcastle University. He has worked extensively on contemporary political philosophy, multiculturalism, toleration, recognition, deontology, human rights, group rights, the nature of liberalism, freedom of belief and expression, democratic theory, international and global justice. Besides numerous articles and chapters, he is the author of *Rights* (Macmillan) and editor of *Group Rights* (Ashgate). His *Essays on Toleration* (ECPR/Rowman & Littlefield) was published in 2018.

Ayhan Kaya is Professor of Politics and Jean Monnet Chair of European Politics of Interculturalism at the Department of International Relations, Istanbul Bilgi University. He is also the Director of European Institute, a Jean Monnet Centre of Excellence. He received his PhD and MA degrees at the University of Warwick, United Kingdom. Some of his latest books are *Turkish Origin Migrants and their Descendants: Hyphenated Identities in Transnational Space* (Palgrave, 2018), *Islam, Migration and Integration: The Age of Securitization* (Palgrave, 2012) and *Europeanization and Tolerance in Turkey* (Palgrave, 2013). His forthcoming book is *Populism and Heritage in Europe: Lost in Unity and Diversity* (Routledge, 2019). Kaya is currently an ERC AdG holder (2019–2024).

Lasse Koefoed is Associate Professor in Social and Cultural Geography at Roskilde University, Denmark, and a highly visible scholar within critical geography. His research interests relate to urban geographies, cities and ethnic minorities, nation and nationalism, postcolonialism, cross-cultural encounters, and everyday life. Lasse has been involved in several major research projects financed by the Danish Research Council for Social Science, including the latest project "Paradoxical Spaces: Encountering the other in public space". He has published widely in international journals like *Mobilities*, *Ethnicities*, *Cities* and *Antipode*.

Ozan Kuyumcuoğlu is a lecturer at the Department of International Relations, Istanbul Bilgi University. He obtained his PhD from Yildiz Technical University in 2018. His dissertation topic is entitled "Viewpoint of Late Ottoman and Early Republican Elite towards Syria". He is recently working on Turkey's Middle East Policy during the interwar period, and on a book project scrutinising the viewpoint of the Kemalists on Nazi Germany.

Sune Laegaard is researcher in Philosophy and deputy head of the Department of Communication and Arts at Roskilde University. He specialises in liberal political philosophy and his recent research focusses on multiculturalism, nationalism, immigration, freedom of speech, toleration, policies of recognition, secularism, and freedom of religion. His numerous recent articles in this field include "Multiculturalism and secularism: Theoretical understandings and possible conflicts", published in *Ethnicities* (2017) and "Burqa Ban, Freedom of Religion and 'Living Together'" in *Human Rights Review* (2015).

Maja de Neergaard completed her PhD at Roskilde University. Her research focuses on 'rural urbanity', problematising the rural/urban dichotomy and exploring practices and lifestyles of urban-rural migrants. She has been involved, as Assistant Professor, on the project "Paradoxical Spaces: Encountering the other in public space" and is currently affiliated to the Department of Urban Studies at Malmö University, Sweden.

Kirsten Simonsen has been Professor in Social and Cultural Geography at Roskilde University since 1996. She is an internationally renowned urban scholar and a leading critical voice in international human geography, a role for which she was recently (2010) awarded an honorary degree by the University of Stockholm. She has published extensively (both journal articles and edited books) in the fields of urban studies, philosophy of geography, space and place, practice theory, minorities, and everyday life. Her latest research projects, all financed by the Danish Research Council for Social Science, are "The Stranger, the City and the Nation" and "Paradoxical Spaces: Encountering the other in public space" (with Lasse Koefoed and Maja de Neergaard).

Dan Swanton is a cultural geographer and a Senior Lecturer in Human Geography at the University of Edinburgh. His work examines the everyday geographies of urban life, with a particular focus on urban multiculture, the geographies of race and racism, and urban ruins. He is particularly interested in developing experimental ways of researching, writing about, and intervening in, the everyday life in cities.

Acknowledgements

The initial ideas that gave birth to this collection were developed at a workshop in Amsterdam in December of 2016, under the title "Spaces of Tolerance: The Politics and Geopolitics of Religious Freedom in Europe", sponsored by the Amsterdam Centre for Contemporary European Studies (ACCESS EUROPE – now the Amsterdam Centre for European Studies, ACES) and its Jean Monnet Centre of Excellence. We would like to thank ACCESS for its generous support, and especially Gijs van der Starre and Jens Kimmel for their administrative assistance without which the workshop could not have happened. It was during that workshop that Peter Jones, Sune Laegaard, and Dan Swanton presented the first versions of their chapters in this volume. We would also like to extend our thanks to the other participants in the Amsterdam workshop, for their comments were crucial to the development of the ideas contained here: in alphabetical order, Avi Astor, John Horton, Yolande Jansen, Sebastiano Maffettone, Domenico Melidoro, Silvia Mocchi, Annelies Moors, Glen Newey, Agnieszka Pasieka, Jonathan Seglow, Lynn Staeheli, Thijl Sunier, and Lorenzo Zucca. The academic discussions at this workshop were also paired with a public event at Amsterdam's cultural centre *De Balie*, entitled "Inter-Spaces: The spaces of encounter between Europeans, new and old". The event brought the workshop's academic participants together with local activists and organisers and the general public, to think together in an informal atmosphere how tolerance can happen in practice. We would like to thank Iris Ruijs and Daphne Dijkman from *PAX*, Bengin Dawod and Fronnie Biesma from *Ondertussen*, and Husam Ibrahim from *Sonono Radio* for their contributions to the discussion, as well as our moderator that evening, Dutch journalist and writer Caroline de Gruyter.

Valentina would like to extend her thanks to the Center Peter Gillies at the University of Antwerp that sponsored the further continuation of activities linked to the completion of this book. She would also like to thank Willem Lemmens, Walter Van Herck, and Patrick Looboyck at the University of Antwerp for inspiring discussions about the subject of this collection. Luiza would like to thank the Centre for Ethics and Global Politics and the Department of Political Science at LUISS Guido Carli University in Rome for a Visiting Professorship in the spring of 2017 that allowed her to work in close collaboration with Valentina on the volume. She would also like to express her gratitude to the Institut für die

Wissenschaften vom Menschen in Vienna for a visiting fellowship in the autumn of 2018 and spring of 2019 that allowed her to complete the final editing tasks in the tranquillity of the IWM – thank you!

Finally, we would like to thank Clive Barnett, the series editor, for encouraging us to submit this collection to the Routledge "Research in Place, Space and Politics" series, and for being open to such an infrequent disciplinary collaboration. Clive's work has been inspirational to us in thinking about the need to bring together normative political philosophy and geographical analysis, so we are especially pleased to be able to publish our book in his series. We would also like to thank Faye Leerink, our commissioning editor at Routledge, and her team, for their precious assistance in preparing the book. Finally, a huge thanks goes to all the authors who agreed to be part of this volume for their inspiring contributions to rethinking the possibilities for tolerance in today's Europe.

Spaces of tolerance

Theories, contested practices, and the question of context

Luiza Bialasiewicz and Valentina Gentile[1]

Across Europe, a variety of parties and movements are re-claiming the rubric of religion in order to draw boundaries between who should, and who should not, belong to the common European home. From right-populist governments in Hungary and Poland, to the *Lega* in Italy and the *Forum voor Democratie* in the Netherlands, the role of supposed 'Christian values' in defining what/who Europe is (and should be) has emerged front and centre (see, among others, Hafez, 2018; Lewicki, 2017; Kalmar, 2018; Krzyzanowski, 2013; Scott, 2018). Over the past decade, calls for a 'reverse crusade' to protect Europeans both against a looming 'Islamic takeover' and 'enforced secularisation' by EU institutions have moved into the public arena from what were previously the extreme fringes of acceptable political discourse (Bialasiewicz, 2006). At the same time, recent rulings of the European Court of Human Rights (Scharffs, 2010; Zucca, 2012; McCrea, 2014) have raised fundamental questions regarding the role of religions in public life and public spaces, and have also opened new theoretical debates regarding the scope of religious freedom (Jones, 2017) and the negotiation of the boundaries of state neutrality and liberal democracy (see, for example, Laborde, 2017; Laborde and Bardon, 2017; Balint, 2017).

Questions of religion have also taken on new dimensions: what have thus far been, in many ways, 'internal' discussions pertaining to the negotiation of religious accommodation and the place of religion in the public realm, have now also become matters of 'external' foreign policy and geopolitics: it is not an exaggeration to affirm that, today, religions are being increasingly framed as both an internal and external security threat. The growing securitisation of religion can be noted, for instance, in European states' monitoring of foreign preachers and the activities of foreign religious organisations. It is even more evident, however, in the management of migration flows: in a number of national contexts, deliberations of migration and asylum are now directly marked by questions regarding the religious background of the migrating non-European Others. Many of these dynamics are, needless to say, not unique to European states: in North America, the increasing securitisation of religion and religious individuals has also been the focus of extensive critical discussion, as too also in other contexts such as the Indian one. Nevertheless, our focus here lies with the distinctively European refractions of such discussions, noting that while they may share a variety of characteristics with other contexts, they are also very much dependent on the uniqueness of European historical and contemporary experiences of the management of religious (and other) difference.

In providing a uniquely interdisciplinary and cross-national perspective on some of the challenges of negotiating the contours of religious tolerance within Europe today, we adopt the notion of 'spaces of tolerance' in both its figurative meaning – as realms of possibility within the liberal democratic order for religious practices and beliefs – but also in its material sense: that is, as the real, physical spaces where these negotiations take place. Focussing on the actual spaces of (and for) tolerance also allows us to cast our gaze beyond the state and the institutional and legal frameworks that are often the privileged focus of such discussions. As noted previously, while our principal focus is on EU states and debates, the volume understands Europe both as a geographical space and as a 'value community' extending beyond the borders of the EU28. We query, in fact, how the value of tolerance as a marker of 'Europeanness' has taken on new salience and has become, in many national contexts, a means of exclusion of 'intolerant others' rather than a principle of inclusion. A number of the chapters in the volume examine this question by looking to how debates over tolerance have become indelibly bound up with debates about migration; indeed, in much of European popular media discourse about migration, questions of the tolerance of difference (religious or otherwise) and 'hospitality' to migrant populations have become entirely intertwined (for a discussion, see Rheindorf and Wodak, 2018; Wodak, Khosravinik and Mral, 2013).

While attentive to this important shift in media and 'commonsense' understandings of 'tolerance' today that have bound it up with a variety of other concerns, the volume also aims to move the discussion of religious tolerance beyond a predominant focus on radical(ised) Islam and to a wider consideration of the place of religion and religious values and ideas. This is in part to counter precisely the sort of entwining of the question of tolerance with the question of how to deal with the arrival (or integration) of religious (read: Muslim) Others in European societies – but also to better understand how to counter the hijacking of a religious rubric (and the discourse of 'tolerance' specifically) by political forces that tolerant clearly are not, as we note in our opening words.

How context matters: tolerance as ideal and grounded practice

Our volume aims to provide an innovative contribution to these very timely discussions by focussing explicitly on the difference that diverse institutional and material spaces make to actual, grounded practices of tolerance. What we wish to add to existing debates, indeed, is a methodological and interdisciplinary reflection on the role of context in making sense of the notion of tolerance both as ideal, and as grounded practice. By putting in dialogue liberal normative theory and an examination of existing legal frameworks together with an assessment of actual practices of tolerance 'in place', we highlight the important ways in which the negotiation of tolerance in practice can inform a principled view of tolerance and vice versa. We ask, in other words, how a certain understanding of equality implicit in liberal democratic institutions 'touches down' in actual contexts, places, and practices – and what happens to such a broader understanding of equality once we properly take into account these latter. The reflection we take forward here thus operates in a double register: aiming to 'ground' ideal liberal democratic theory and, at the same time, to bring normative reflection to the sort of 'grounded theory' privileged by critical human geography.

The slippery relation between religion and politics has long been a concern of liberal democratic theory. Over the past couple of decades, this relationship has also come under scrutiny by a much wider body of scholarship, including post-colonial and decolonial critiques, as European liberal democracies struggle to find answers to managing the complexities of multicultural, multinational and multireligious societies. Nevertheless, examining the mainstream of liberal normative conception of tolerance that has characterised Anglo-American political philosophy of the last three decades and that has strongly influenced European normative thinking (for example, Forst, 2013), the role of context appears quite limited. With reference to this, two distinct, realist critiques have been raised: tolerance-as-idealisation and tolerance-as-power. We shall focus on the first critique here, and will consider the second, tolerance-as-power, in the subsequent section of the chapter. The first critique focuses on liberal ideal-theorising envisaging a 'realist turn' (Galston, 2010, p. 386) in liberal democratic theory, as it argues for a clearer distinction between the political and the moral realm. 'Political realism', following Bernard Williams (2005, p. 3) is seen, indeed, as an antidote to a highly idealised and unfeasible version of liberal egalitarianism, and especially to the approaches of Rawls and Dworkin (Galston, 2010, p. 387ff; Gray, 2000). Indeed, 'political realism' has attempted to provide a properly 'contingent' response to the deep moral pluralism that characterises contemporary democracies (Gentile, 2018) and, therefore, to respond to the special demands for protection articulated by citizens of faith. As Williams argues:

> The case of toleration is, unsurprisingly, a central one for distinguishing between a strongly moralized conception of liberalism as based on ideal of individual autonomy, and a more sceptical, historically alert, politically direct conception of it as the best hope for humanly acceptable legitimate government under modern conditions.
>
> (Williams, 2005, p. 138)

The emphasis on context plays a central role in this critique, as it suggests that only a context-sensitive approach can be a legitimate source of political normativity (Rossi and Sleat, 2014, p. 694). We aim to engage directly with this approach, affirming that the rethinking of tolerance and its (ideal) philosophical justification cannot be isolated from actual practices and contexts. What is more, we seek to show that an ideal liberal egalitarian view of tolerance need not to be in contrast with actual practices of tolerance, and an appreciation of historical as well as socio-cultural 'contextual' differences. If there is, in fact, one key contribution that we hope to make with this volume, it is to show how these two approaches might help to reinforce each other.

Building on political realist critique, we might distinguish three different ways of approaching the idea of tolerance. First, **tolerance as a practice**: this view emphasises the eminently subjective character of the justification of coercion and the impossibility to attain one shared view of tolerance. On this account, to look at actual practices and modes of coexistence is the "best hope for humanly acceptable legitimate government" (Williams, 2005). However, this view is neither stable, nor does it provide a sufficiently strong defence of rights and freedoms, including religious freedom. Second, **tolerance as an ethical doctrine**: this view implies that there exists an objective and somehow practice-independent way to justify liberal tolerance – for instance, a view that justifies tolerance on the basis of an over-demanding (moral) value of autonomy, such as Kantian autonomy. On this point, we agree with realist scholars (and Williams in particular) that appeals to such an independent ethic are not only unfruitful but also undesirable as far as they promote a form of liberalism that risks becoming yet "another sectarian doctrine" (Williams, 2005, p. 130). Finally, **tolerance as political justice**: this view is chiefly intersubjective as far as it appeals to a form of justification that involves reasons that we reasonably believe others cannot reasonably reject (Rawls, 1996). This view might be said to be 'practice-dependent' in two important ways.[2] First, because its justificatory structure reflects liberal democratic institutions and, especially, a distinctive institutional understanding of citizens envisioned as free and equal. Second – and even more pertinent to the aims of this collection – because an institutional practice-dependent view of tolerance[3] is necessarily mediated by contextual, historical, and social conditions. While it is institutional and 'idealised' ways of thinking and justifying tolerance that secure fairness and impartiality, it is also non-strictly public values and identities that emerge 'in place' and at the level of civil society that significantly contribute to the development of those civic and political virtues necessary for the stability of European democracies. Tolerance as 'political justice' is, therefore, always profoundly context-sensitive.

The importance of context in shaping the conditions of social and political life has long been a guiding preoccupation of contemporary human geography, indeed in many ways marking its very raison d'être. There is a long disciplinary history that accounts for this, including a reaction against the dominant quantitative and economistic models of the 1950s and 1960s. The emergence of context-sensitive approaches can be already traced to the humanistic geography of the late 1970s and 1980s, and subsequently to the emergence of the critical and radical human geography of the 1980s and 1990s, inspired in its various strands by Marxist and post-Marxist theory but also by poststructuralist approaches (for a comprehensive overview, see Gregory et al., 2009). What is striking, however, is that while a great part of critical human geography of the past (by now four) decades has been strongly focussed on questions of social and spatial justice, an explicit engagement with normative questions regarding equality, justice, rights, and, pertinent to the concerns of this book, tolerance, has been largely absent, as Barnett (2012, 2017, 2018) has argued on a number of occasions.

Writing in a seminal essay on one of the leading geography journals 20 years ago entitled "Ethics Unbound: For a Normative Turn in Social Theory", Andrew Sayer and Michael Storper (1997, p. 1) noted how:

> critical social scientists have been coy about talking about values. They frequently use negative terms such as 'racist', or positive terms such as 'democracy', which carry a strong evaluative message, but there is often a refusal to present the arguments for the evaluations. Indeed even to ask for such justifications is likely to be taken as shocking and threatening – as implying the acceptability of the thing being opposed – rather than just a request for clarity about what exactly we oppose or favour and why and with what implications. [. . .] One of the great ironies of postwar social science is that, although Marxists and related radicals criticised liberal positive social theory as reactionary and apologetic, liberalism had a long-standing and far more developed normative side [. . .]. By contrast the normative basis of Marxist-influenced radicalism was under-examined and flimsy. Outside the disciplines of politics and philosophy, most social scientists had little or no acquaintance with normative theory.

Sayer and Storper note that while the late 1990s brought, for the first time, a shift in this respect, the engagement of geographers and other spatial scientists with normative questions remained relatively constrained. Reviewing the field a decade later, Jeff Popke (2006, p. 510) similarly invoked human geographers to "a more direct engagement with theories of ethics and responsibility", noting that it was not sufficient for geographers to "reclaim the social as a site of geographical analysis" (which they certainly had done over the preceding decades) without also "re-inscribing the social as a site of ethics and responsibility" and thus engaging fully with (also) normative visions of the social. It is interesting – and directly relevant to the discussions in the present volume – that the move to a fuller engagement with normative questions (including questions of tolerance) came with geographers' growing attention to questions of migration, difference and postcoloniality. Cities became a privileged site of such discussions starting from the early 2000s, also often in direct conversation with attempts by policy makers to find new ways of institutionally managing difference in urban space, whether religious or otherwise. The work of Doreen Massey (1994, 2005) was fundamental in this regard, offering new ways of understanding relations of difference in space in a more open and 'relational' fashion. Similarly, Ash Amin (2004, 2006) suggested a new reading of urban polities through a 'politics of propinquity' based on the constant negotiation of relations with others: as he termed it "negotiating the immanent effects of geographical juxtaposition" (2004, p. 39). Both Massey and Amin called upon geographers to be attentive to the multiple relational networks making up cities that also 'stretch' urban relations (and politics) of responsibility: not just to those most proximate and already-here but also to distant others (that which Massey termed in her seminal 1994 book "a global sense of place"). It is through these discussions, strongly focused on the urban, that normative and ethical questions began to enter the preoccupations of human geographers. Work inspired specifically by "right to the city" approaches has also provided important impetus in bringing together concerns with context and normative theorising (Harvey, 1976/2009; Mitchell, 2003). But as important and influential as these two strands of theorising have been in bringing an engagement with ethical and normative questions to geography, a critical discussion of the normative bases of some of their assumptions and propositions remains under-developed. Writing in a recent review, Barnett (2018, p. 319) has bemoaned, indeed, that much of critical spatial theory guided by the "right to the city" approach "actually retains many of the most problematic features associated with normative philosophies of ideal justice", including insufficient attention to, precisely, the *contexts* of ideal theories of justice, rights, tolerance, and values.

The question of context is also highly relevant to the ways in which geographers have approached the study of religion more specifically. As Dittmer (2007, p. 737) noted in one of the first reviews on the topic, the contribution that geographers could provide to studies of religion and religious practice lay precisely in approaching religion "as an unstable, contextualized signifier", thus "shifting focus from the object of religion to the subjects who contextualize it" (see also Ivakhiv, 2006). The most recent work on religion in geography has indeed focussed on place-based negotiations of religious practice (such as the collection by Kong, Olson, and Hopkins, 2013; see also the review by Tse, 2014) and on the everyday materialities of religion (Gilbert et al., 2018). With this volume, we hope to add to such work by putting geographical reflection in direct dialogue with normative discussions in political philosophy, hoping to answer in part some of the questions raised by Sayer and Storper 20 years ago, querying "how much of the traditional arguments of ethics and political philosophy can be salvaged from the fundamental critiques of recent years? And what difference is improving the sociospatial [. . .] awareness of normative theory likely to make to its conclusions?" (1997, p. 15) – in our case, regarding conceptions and practices of tolerance specifically. As the authors write:

> If ethical principles have to relate to particular kinds of social organisation then there is a sense in which they must also have histories and geographies. Indeed, geographers have shown that social organisation involves not only social interdependencies, but spatial ones as well, including relations of proximity and ordered interplace relations of exchange at a distance. [. . .] Actual geographies of these forms of social organisation have always been complex, involving patterns of mixing and exclusion, boundaries and hierarchies that reflect economic, cultural, political, and technological influences. It is doubtful that there was ever an original state in which many of these relationships had a geographical form which was tidily segregated and simply organised and the forces of modernity continue to stretch and churn them up. [What is more] the embedding of actions and their consequences in space produces

territories which are a palimpsest of past rights and wrongs which continue to affect the lives of subsequent generations [. . .] posing problems of justice which more abstract discussions that ignore spatial settings and temporal change fail to address.

Today's discussions on religious tolerance in Europe must therefore be embedded in such longer histories and geographies to fully appreciate the ways in which they are both (ideally) understood and imagined, and negotiated (in practice) in specific contexts. To quote Sayer and Storper (1997, p. 11) once again, "discussions of ethics, rights, and the good [. . .] must all ultimately involve what people living in specific material contexts, do with and to others, and what others can do with and to them".

Managing religion, managing difference?

Another way in which we hope to put not just tolerance but also religion more broadly 'in context' is by drawing attention to the ways in which debates about religious freedom/religious tolerance have become a foil for other debates and negotiations and how intolerance itself is deeply intersectional. As we remarked at the outset, in today's Europe, intolerance of religion (and of Islam in particular) has become a stand-in for other fears of difference, often strongly racialised, as in public and media discourse it is more acceptable to speak of the 'challenges' of religious difference for European societies rather than speaking directly of race or other identitary fears (see Wodak, 2015; Wodak, Khosravinik and Mral, 2013).

Wendy Brown, whose work on tolerance has been perhaps most influential for contemporary geographers has argued, in fact, that the 'work' that

> tolerance does today for liberalism is different from what it has done in the past. Liberalism has two specific problems today: on the one hand, it's handling complexly multicultural orders [. . .] and it's representing the supremacy of Western civilization. [. . .] Tolerance is playing a really crucial role in both. In the first, I think that tolerance is operating as a supplement to equality, not as an equivalent but as a supplement in the Derridean sense. It is finessing the differences that equality itself cannot manage, can't realize in liberal democracies. More specifically, if equality in liberalism pertains to sameness, tolerance always pertains to difference. So tolerance is a crucial supplement to equality in liberal democracies that understand themselves as suffused with difference and not only sameness. Liberal democracy has always promised equality on the basis of sameness; if what we now have is articulated and even antagonistically articulated differences, tolerance is the supplement that is managing those differences apart from equality. [. . .] And then, on a civilizational level, tolerance [. . .] is identified as a site of Western supremacy, or a site of supremacy vis-à-vis the imagined intolerant.
>
> (Brown and Forst, 2014, pp. 60–61)

We cite Brown's argument at length here for it nicely highlights precisely the ways in which notions of tolerance have become a mode of managing (and indeed securitising) religious and other difference in contemporary Europe. A great number of scholars have described the ways in which discourses of tolerance have taken a prominent role not just in abstract discussions over the proper place of religious difference in European societies but also in policies designed to manage it, most directly through the management of migration, integration and citizenship regimes.

In her work on the interpenetrations of race and Islamophobia in contemporary Europe, Lewicki (2018) notes the ways in which these two have become indelibly interwoven in German citizenship regimes. Citing the work of Sayyid and Vakil (2010, p. 13) she highlights how the racialisation of bodies "was never exclusively focused on visual markers: biology was 'marked at the same time as religion, culture, history and territories were marked and used to group socially fabricated distinctions between Europeanness and non-Europeanness'". She argues that today's migration and citizenship regimes have acted to further demarcate (and harden) racialized ethnic, national, and religious categories, with religion growingly becoming a key discriminant in delimiting access to the European space of rights:

> once Germany had been identified as a harbour for the perpetrators of 9/11, the government's regulatory attention turned to Germany's migrant populations' religious profile [. . .] religion was [turned] into a salient marker of racialized alterity. The conditions of entry to the 'community of value' were specified further in the 2000s. Participation in integration courses was made a mandatory requirement for citizenship acquisition, and naturalization examinations were introduced.
>
> The national examination, supported by preparatory tools, such as textbooks and mock examinations, assessed, among other things, knowledge of constitutional norms that regulated sexuality and gender. This test, therefore, as Schirin Amir-Moazami notes, only aimed ultimately to disqualify 'disloyal' and 'illiberal' applicants but actually was designed to educate what were framed as less-'civilized' populations. By pitting deviant outsiders against an in-group who allegedly held up liberal sexual and gender norms by default, citizenship tests reflected and produced hierarchical binaries. This and a much wider array of integration policy initiatives further sharpened the racialized boundaries of the nation and added a specific focus on Muslim religiosity to the regulatory framework.
>
> (Lewicki, 2018, p. 12)

Germany is of course in no way unique in this shift: a growing body of scholarship on what Duyvendak, Geschiere and Tonkens (2016) have termed the 'culturalisation of citizenship' has examined how, in contexts ranging from Scandinavia to the Netherlands, appeals to 'tolerance' and 'liberal orientations' towards sexual and gender rights, in particular, have increasingly been used by European states as a means to rationalise restrictive immigration policies and other coercive practices against ethnic and religious others. To cite Eric Fassin (2010), "sexual democracy [has become] the language of national identity throughout Europe – in a common context of anti-immigration backlash, 'gender equality and sexual liberation' provide a litmus test for the selection and integration of immigrants, in particular from the Muslim World". The Dutch context is in many ways exemplary in its bringing of questions of 'proper tolerance' directly into its migration and citizenship regimes, framed through four specific cultural tropes: "gender equality, sexual freedom, freedom of speech and individualism" (de Leeuw and van Wichelen, 2012). It is precisely these tropes that form the basis of the Dutch Integration Exam. The exam 'package' includes a one and a half hour film entitled *Naar Nederland* that applicants are required to watch, meant to grant an overview of "Dutch society, values and culture". The film "is guided by a female presenter who seems to stand for both the image of the modern Dutch woman and, through her narrative, the voice of the state that speaks to its new immigrants. The narrator lists for the viewer a number of offences targeting women as strictly punishable by Dutch law – from honour killings, to female circumcision and domestic violence – denoting, as de Leeuw and van Wichelen remark, "violence against women as practices that they do (murder, domestic violence, bodily aggression) and informing the viewer that they are punishable by law in our society [. . .] suggesting that these modes of religiously condoned gender-based violence are not similarly evident in the Netherlands" (2012, p. 198). The film had also attracted international attention because of its inclusion of scenes of two gay men kissing, as well as those of nude sunbathers (subsequently removed): both used to ostensibly 'assess' the tolerance of the exam takers (see the discussion in Bracke, 2011, 2012).

In these and similar cases, the 'assessment' of the 'liberal tolerance' of migrant others by the institutions of the state is, as Brown (2006; Brown and Forst, 2014) suggests, nothing else than a more or less exclusionary mode of 'multicultural governance'. But can we identify instances where liberal approaches to the management of religious difference can be something other than simply regulatory instruments of the state, or modes of exclusion tinged with Western civilizational supremacy? Here, we return to the realist critiques of liberal tolerance. The second critique, that which we termed the 'tolerance-as-power' approach, sees the liberal conception as reflecting an indiscriminate and pervasive strategy of power, one that operates at the interstices of "power differentials" (Brown and Forst, 2014, p. 35) and that contributes to inscribe a divide between a tolerant 'us' and intolerant 'others'. An examination of the actual contexts and practices of (what is presented as) tolerance reveals here the character of liberal tolerance as "a form of multicultural governmentality and Western civilizational discourse" (2014, p. 20).

We must be careful, however, to distinguish in this discussion between two different conceptions of tolerance. One modern version of tolerance, which emerged in the aftermath of the religious wars in Europe, might be called a "permission conception" (see Forst, 2013, p. 518; Brown and Forst, 2014), or in Rawls' words (1996, p. 157ff) "modus vivendi". The second conception is instead focussed on the form of respect we owe to each other as citizens: this is what Forst (2013, p. 158ff);

has termed the "respect conception". Contemporary liberal theory is generally concerned with the second type of conception of tolerance. The above distinction is important for this is precisely the focus of the first critique of tolerance we identified earlier. Thus, whereas the first critique recalls contexts and 'actual practices' to suggest a return to the first, more realist conception of tolerance (see especially Gray, 2000 and McCabe, 2010), tolerance-as-power, denies the relevance of this distinction as far as it reduces the very idea of tolerance to simply reproducing existing hierarchies of power and exclusion. In her exchange with Rainer Forst, Brown (2014, p. 19) asks, indeed:

> what happens when tolerance shifts from belief as its object to identity as its object, from religion to race? [. . .] How, also, does tolerance substitute for equality while purporting to support equality? How does tolerance subtly stratify and abject certain peoples who have formal rights and equality?

These are all questions, she notes, that mark the 'work' done by tolerance today as, predominantly, a mode of managing difference – and, as we note above, a powerful mode of multicultural governance.

How, then, should normative liberal theorising respond to this second critique? We believe that it rightly forces liberal scholars to rethink the nexus between equality and tolerance, neutrality, and diversity – especially when religious voices and identities are articulated in ways that Western secular understandings might willingly or unwillingly exclude and silence. We contend, however, that it is possible to rescue the idea of **tolerance as 'political justice'** from this critique. If it is true that tolerance as 'political justice' is not 'power-free', to cite Brown (Brown and Forst, 2014, p. 42), *this does not mean that there is no room for distinguishing what is normative and therefore 'just' in a liberal sense, from what is normative in a Foucauldian sense but 'not-just' from a liberal perspective* (on this, see also Laborde, 2017, p. 37). We believe that there are different modes of living with and in difference, just as there are different ways in which the boundaries of state and religion might be constructed, negotiated, and justified. As several contributions to this book show, the problem is not so much that of getting rid of forms of coercive power but rather that of distinguishing between conceptions and practices of tolerance, neutrality or 'conviviality' that are 'better', from those that are 'worse' (see Nancy Fraser's well-known critique of Foucault, 1981, p. 286). Thus, a secular society might rightly choose to refrain from publicly endorsing certain religious symbols or practices for the sake of equality. At the same time, by endorsing forms of 'militant comprehensive' secularism it can put at risk the very idea of equality on which a just and tolerant society should rest (see both Laegaard and Galassini in this volume). So, while it might be already problematic, for example, to justify banning the headscarf for Muslim women if they are teachers in public schools, it is even more problematic to extend such a ban to all public services and spaces (see Gentile in this volume). This becomes even more unjustifiable when other religious symbols – such as the crucifix – are, instead, allowed (see on this in Brown and Forst, 2014, p. 44). At the same time, it should be possible to find a right balance between a genuine protection of citizens' beliefs and religious practices compatible with ensuring an equality of status, power, and relations amongst *all* citizens. Forst has clearly outlined this problem as follows "We have to have a notion of democratic justice by which we provide certain criteria of what a good reason is in order to establish or reform an institution and to determine how to interpret basic rights" (Brown and Forst, 2014, p. 30).

Through the dialogue presented in this collection, we show that a more careful reading of contemporary liberal egalitarianism suggests that tolerance can still serve as a useful tool in providing answers both to the challenges of realist criticisms as well as to the 'on-the-ground' challenges of contemporary European liberal democracies. We believe that by bringing together normative theory with a context-sensitive analysis of particular places where questions of tolerance are being negotiated 'in practice', we provide an original contribution to ongoing discussions surrounding the appropriate place for equality and inclusion in today's European societies. Accordingly, we have chosen to structure the book in three intersecting parts, aiming to address the aforementioned challenge. Part 1 provides four different perspectives on the normative ideal of tolerance and the negotiation of religious difference in contemporary liberal democracies; Part 2 focuses on what are the limits of tolerance, examining the securing of tolerance but also its increasing limitation through processes of securitisation; Part 3 closes the volume by offering three different urban case studies, focussing on the actual 'spaces of tolerance' (and intolerance) and the ways in which these are negotiated on the ground.

Negotiating freedom and religion: tolerance, neutrality, conviviality

This opening part of the volume is intended to set the theoretical and conceptual context for what we term **tolerance as 'political justice'**. This is the only part of the book comprising four chapters: bringing together four different contributions allows us to adequately address realist critiques but also to put into dialogue normative political philosophy accounts of tolerance/neutrality with perspectives from geography. The four chapters are thus intended as a series of windows into some of the diverse ways in which the religion-state relationship – and the nexus between freedom, equality, and religion – has been understood, negotiated, and justified.

Both of the realist critiques that we outlined previously wonder whether liberal tolerance is able to take seriously the forms of pluralism that characterise 'actual' democracies, especially when disagreements about the boundaries of the 'good life' necessarily translate into disagreements about the very content of justice. The ideal of tolerance thus risks, as we already noted, being just "another sectarian doctrine" (Williams, 2005). Or, it risks proving inadequate to accommodate religious citizens' claims. On the other hand, tensions between protecting freedoms (and especially religious freedom) and securing equality are not simply theoretical – they emerge with full force in today's European societies.

How should a view of tolerance as 'political justice' be framed to deal with the kind of pluralism that characterises contemporary European liberal democracies? The first two contributions provide different, though complementary, normative answers to the problem of engaging difference: one focussed on the political value of tolerance and its relationship with the ideals of democratic equality and political autonomy, the other distinguishing a specific account of neutrality, namely 'active indifference', from different modes of 'managing diversity', in particular with regards to multiculturalist approaches and the politics of difference.

Gentile's opening chapter clarifies the relationship between two central tenets of liberalism: a robust protection of citizens' religious freedom, and a political value of tolerance that secures the normative stability of liberal institutions. While liberalism suggests that these two tenets are compatible, there is considerable tension between them. Gentile maintains that these two can work in harmony on the basis of a fundamental commitment to securing citizens' political autonomy. The chapter notes how the standard approach, based on the ideal of 'equal liberty' through which liberal egalitarianism has dealt with this issue (see especially, Eisgruber and Sager, 2007), faces the problem of disregarding what is really at stake with religious freedom. Gentile argues that the determination of the scope of religious freedom is a problem of non-comparative justice, one that guarantees (equal) access to a constitutional right. Nevertheless, this is still a problem of egalitarian justice as far as it asks to determine the most appropriate view of citizens' equality which helps to single out legitimate claims of religious integrity. The chapter thus suggests that Rawls' political value of autonomy (1996) and the related notion of democratic equality can significantly help in determining the scope of religious freedom in contemporary democracies. Rawls' public justification and its appeal to political autonomy displays a genuine liberal commitment to citizens' religious integrity. The political value of integrity is meant to determine the appropriate constraints to religious freedom in contemporary democracies.

Peter Balint's contribution in chapter two defends a specific account of 'neutrality' as the most valuable form of tolerance. The idea of neutrality has often been the target of critique by multicultural scholars and, more generally, by supporters of the ideal of respect for differences who commonly see this ideal as incapable of accommodating diversity. In responding to such critiques, Balint argues that while neutrality is unrealisable, it can still constitute a valuable "action-guiding political ideal". He shows that there are two ways in which a difference-sensitive approach can be realised in existing democracies: either by removing preferential support for *any* form of difference, or by actively assisting *all* forms of difference. Balint opts for the first approach and argues that 'neutrality', understood as form of 'active indifference', ideally involves "withdrawing support for favoured ways of life", rather than actively recognising the different ways of life of its citizens.

The chapter that follows, Peter Jones' analysis of tolerance and toleration, allows us to begin to engage, at least in part, Wendy Brown's critique of the normative conception of tolerance. As we noted previously, Brown emphasises the difficulties of disentangling an idea of tolerance, understood as a form of reluctant acceptance of what we usually object to, from a more positive account, one able to promote and value differences of identity. Her concern lies, indeed, with "what happens when tolerance shifts from belief as its object to identity as its object, from religion to race?" The third chapter in this part engages directly with this challenge and offers a defence of the ideal of 'tolerance' capable of responding adequately to forms of difference of identity emerging in plural European societies. In his contribution, Jones distinguishes an idea of toleration, which entails "objecting to but not preventing whatever it is that we tolerate", from an ideal of 'tolerance' which entails "abstaining from disapproval or dislike". Jones argues that a society in which identitary differences become sources of hostility and conflict, mutual toleration amongst its citizens may well be a condition to which it should aim. Toleration will, then, be an instrument for coping with non-ideal circumstances, but it cannot represent the condition that should prevail in an ideal society (that is marked by identitary differences).

In the chapter that concludes this part of the volume, Dan Swanton engages the work of Brown directly and argues for the need to 'side-step' the conceptual terrains of tolerance engaged by the previous contributions in order to set abstract-ideal discussions of tolerance also alongside the 'messy lived realities' of everyday spaces and practices of tolerance. He confronts both normative-ideal readings of tolerance as well as those conceiving tolerance as a discourse of power and governmentality with what Back and Sinha (2016) refer to as a 'situated ethics of conviviality': an approach, Swanton argues, better able to capture not just the varied lived forms of tolerance and conviviality, but also other affective orientations to difference that may or may not be tolerant or convivial. Through the discussion of two specific urban instances, Swanton's chapter reflects on how the lens of what he terms 'infrastructures for living with difference' (infrastructures that can be both material as well as ideal and symbolic, fleeting or longer-lasting) can allow us to better understand the workings of such situated ethics, and the varied ways in which differences encounter 'in place'.

Securing and securitising religious tolerance

"Religion is both an object of processes of securitization and of toleration" Laegaard suggests at the outset of his chapter that opens the second part of the book. This part picks up on the question of the bounding of the philosophical 'spaces for tolerance' discussed in the first, but focusses more specifically on the different ways in which religion has been made the object of securitisation and, especially, the tensions between securing (and making possible) religious tolerance, and the growing securitisation of religion. The securitisation of religion in Europe can be traced back to already the early 1990s, with the emergence in popular and political discourse of a variety of 'civilisational threats' (in many instances, directly linked to particular religious identities). Nevertheless, the imagined nexus between religious difference and national and international (in)security emerged with full force in public debates with the combined effects of terror attacks in European cities and the 'hospitality crisis' of 2015–2016. In the political discussions that followed, the question of European security became indelibly bound with the question of how to respond to religious difference: religious difference that now became coded directly as 'danger'.

In these debates, questions of European states' internal and external security became one and the same: whether in the deliberations of how to deal with the management of migration and the securing of borders, or in facing the threat posed by the trans-national networks of radical Islam. The chapters in this part highlight some of the ways in which the spaces of religion and 'religious individuals' have become increasingly securitised. While the first two chapters (Laegard and Galassini) focus predominantly on the tensions between tolerance and the risks connected with securitising religions within European states, the last one turns the attention to the problems connected with 'securing' tolerance and hospitality at the borders of Europe and, more precisely, with reference to Turkish state's reception of Syrian refugees.

In her critique of tolerance, Brown (Brown and Forst, 2014, pp. 18–19) emphasises how certain values, including "individualism, secularism, enlightenment and civility", are all connected to an ideal of 'tolerance' that reflects a Western liberal civilisation as opposed to intolerant Others, most notably, Islam. Through these discourses, Western democracies justify the securitisation of religions and the denial of rights in the name of tolerance. Here, Brown's critique of tolerance-as-power becomes clear. She asks, "I am interested in how tolerance was deployed in the years 2001 through 2004 to justify the invasions by the US and Britain of Afghanistan and Iraq" (Brown and Forst, 2014, p. 19). In responding to this challenge, Forst argues that tolerance needs a third component beyond objection and acceptance to work, namely the rejection component (Brown and Forst, 2014, pp. 23–24). The rejection comes into play when we are in front of beliefs, practices, and behaviours that simply cannot be tolerated. It is about the limits of tolerance. Brown seems to fear that these limits are already implicit in what she sees as a hegemonic "Western liberal civilizational discourse".

In the previous paragraphs we argued that the problem for normative philosophy is not that of providing a 'power-free' view of tolerance: rather, it is that of offering a reasonably justifiable conception. However, even rescuing liberal tolerance from Brown's accusation one problem remains: if liberal tolerance does not accommodate different standards of value and worldviews (even if incompatible with liberalism), it becomes an intolerant doctrine itself. Here, the problem arises. If, on the one hand, it seems to be contradictory for 'tolerance' to require intolerance toward different practices or beliefs, on the other hand, to what extent should liberal 'tolerance' require the tolerance of non-liberal objectionable practices? Where should the line between the rejection and objection component be drawn?

These problems are addressed in Chapter 5 by Laegaard. His contribution focusses precisely on the limits of religious tolerance and emphasises the risks connected with the securitisation of religions. Laegaard argues that this kind of securitisation "is often linked to threats from abroad, e.g. in relation to religiously defined immigrants or religiously motivated terrorism". In recent years, religions and especially Islam have been the target of processes of securitisation in several European societies. Laegaard shows how these processes of securitisation of religion are likely to modify the meaning and scope of religious tolerance and rights protection. The appeal to neutrality that once was adopted for "handling religious differences in a legitimate way" becomes now an instrument for singling out and targeting specific religious groups and practices. Thus, liberal societies appear redesigned and divided into 'first-order' citizens, whose practices are approved, and 'second-order', 'illiberal' citizens, whose practices and beliefs are rejected. The space for toleration of disapproved practices disappears and is replaced, Laegaard suggests, by a "comprehensive liberal interventionism and militancy vis-à-vis religion".

The chapter that follows also provides a critique to the form of comprehensive 'militant secularism' recently adopted by European societies and institutions, and especially by the European Court of Human Rights, in dealing with citizens' religious claims. Margherita Galassini argues that the ECHR's approach to religion is one of 'militant secularism' and urges a more moderate, context-sensitive way to deal with religious diversity. She sets out an account of tolerance termed 'tolerant pluralism' which should allow European societies to pursue their own "reasonable interpretation of the relationship between state and religion" in harmony with accommodating religious diversity.

Kaya and Kuyumcuoğlu's chapter closes this second part with a reflection on shifting popular and governmental attitudes towards Syrian refugees in today's Turkey. Examining both historical sources and the discourses of contemporary political leaders, they describe how questions of tolerance and, specifically, hospitality towards Syrian refugees as 'guests', are framed with appeals to the past. They highlight how the 'securing' of hospitable relations between Turkish 'locals' and Syrians in the present is reliant upon distinct understandings of shared histories of interaction and co-presence, as well as narratives of a shared 'cultural intimacy' (Herzfeld, 2016). The authors employ this notion to analyse how feelings of 'alikeness' and produced and reproduced not just through state rhetoric but also, importantly, through everyday acts, attitudes, and those that Herzfeld terms forms of 'iconicity' (Herzfeld, 2016, p. 93). By focussing on both the changing discourses and legal frameworks of the Turkish state, as well as on the attitudes of local citizens (by now sharing space with Syrians for several years in many Turkish cities), Kaya and Kuyumcuoğlu draw attention to the ways in which elite framings of the 'tolerance and benevolence' of the Turkish state interact with often quite disparate quotidian experiences and orientations, of both Turks and Syrians. They also note, importantly for the broader aims of this volume, how questions of tolerance/intolerance of – and in – Europe are very differently perceived by Syrian refugees in Turkey than European public opinion might imagine. Indeed, studies cited by the chapter note how for a majority of Syrians currently resident in Turkey, Europe is seen as a space of intolerance and inhospitality.

Everyday spaces of tolerance

With its focus on both the legislative and everyday spaces of tolerance, Kaya and Kuyumcuoğlu's chapter neatly leads into the third and last part of the book, focussed precisely on the various ways in which 'actual' practices of religious tolerance, and also hospitality and solidarity, are negotiated on the ground, in the 'real' spaces of European cities. Over the past years, contests over material spaces have become increasingly crucial to negotiations over the (also physical) boundaries to the tolerance of religious practices in Europe. Most prominent among these have been the highly publicised conflicts over the building of mosques, or the banning of particular forms of mosque architecture like minarets (the object of a referendum in Switzerland, and of proposed referenda in both Germany and Austria, as well as a variety of local and regional ordinances in Italy – for a discussion, see Bialasiewicz, 2017). Nevertheless, contested negotiations over the tolerance of the religious practices, sites, and sounds of non-European Others have certainly not been limited to formal and visible spaces of Muslim worship.

The three chapters in this part focus on three different European cities (Copenhagen, Ghent, and Rome), in three very different national contexts: national contexts distinguished by quite different histories of the role of religion in urban and public life, different attitudes (both public and popular) to the negotiation/acceptance of religious difference but also different histories of migration. These three 'case studies' that close the volume allow us to highlight, in part at least, how negotiations over religious freedom actually 'touch-down' in specific localities. They also allow us to foreground how what are, ostensibly, negotiations about religious difference are indeed often about something else altogether: that is, how religious difference and debates surrounding its proper 'tolerance' are, more often than not, a foil for wider debates regarding the very place and right-to-space of non-European Others in European cities.

The opening chapter of this part, by Simonsen, de Neergaard, and Koefoed, engages directly with this problematic in the context of Copenhagen. The authors begin their analysis by describing the shifts in the Danish public debate about Islam, focussing in particular on the contradictory politics of visibility with regards to the presence of Muslim 'Others', embodied as well as architectural. The chapter takes a phenomenological approach and draws also on post-colonial theories of whiteness to analyse encounters with Muslim difference, noting how relations of (in)visibility are crucial to understanding how difference 'comes into the public'. The authors provide an in-depth discussion of the popular and media reception of a new purpose-built mosque in Copenhagen in 2014, noting how frames of stigmatisation and securitisation were invoked to present the mosque not just as an 'enemy of tolerance' but also a potential security risk: "bringing into the Danish cityscape not just Islam but also the war on terror". Drawing on their fieldwork, Koefoed, de Neergaard and Simonsen note, nevertheless, 'cracks' in the exclusionary and fearful public discourse: both through the (highly visible) presence of leaders of other religious communities in supporting the opening of the mosque but also in the reactions of local inhabitants interviewed by the authors, highlighting in this fashion some of the 'paradoxes' of religious visibility and its emplaced politics.

The subsequent chapter, Luce Beeckmans' analysis focussed on the 'religious place-making' strategies of an Afro-Christian church in Ghent, extends a number of the questions raised by the authors of Chapter 8. Beeckmans' chapter focusses strongly, indeed, on the strategies of selective visibilisation adopted by the Redeemed Christian Church of God (RCCG), and on how such strategies are closely bound to an ongoing negotiation of the right to migrant religious presence. She notes how such selective visibilisation is evident in the church's choice of physical location on the outskirts of the city, 'disappearing' into the surrounding landscape within an ordinary terraced house (and within what appears to be a 'private' space) – while at the same time, highly 'public' in its concern not just to be seen as abiding by all local rules and regulations but also as a 'positive' actor in the integration of migrants in the city. This second point underlines another important aspect of Beeckmans' case study: as the chapter highlights, religious actors (and spaces) like the RCCG in Ghent are often 'tolerated' also because they (at least partially) fulfil some of the roles of the local state in providing a sort of first-point-of-call 'arrival infrastructure' for migrants, offering not just a space for congregation and the exchange of crucial information but also various other forms of mutual assistance and solidarity.

The final chapter in the volume, Bialasiewicz and Haynes' discussion of the role of religious organisations in migrant assistance in Rome, also probes the changing role of religious actors today in a wider political economy of 'humanitarian assistance'. They examine, in particular, the role of Catholic organisations and their discourses and practices of 'hospitality' to migrants as a counter to the increasingly exclusionary policies of the Italian state. Focussing on Rome's central train station, Termini, the chapter outlines how the public spaces of this key transit site have been made progressively more inhospitable to its growing homeless population, both migrant and non, and how these transformations reflect other wider shifts in the physical as well as legislative exclusion of irregular migrants from access to basic rights. Looking to the role of two religious organisations active in providing daily assistance to migrants in and around Termini station, Bialasiewicz and Haynes note the precarious nature of such help. They note that while these organisations' provision of momentary 'hospitality' may provide relief for some of the migrants' most urgent needs, the high visibility of such forms of assistance risks not just to assuage the conscience (and obligations) of local authorities but also to over-expose to the public gaze the presence of homeless migrants and their condition, opening the door to episodes of intolerance, if not direct violent confrontation: episodes, as they note in closing, that have become increasingly common in the current Italian political context.

Negotiations of questions of visibility and presence in public space, and of access to basic rights, are directly pertinent to our discussion of the boundaries of tolerance in today's Europe. Such emplaced, 'grounded' negotiations matter for they speak to how religions and religious difference are encountered in specific places. They matter in a European context where religion and religious difference are increasingly not just securitised, but made the object of violent attacks. To our mind, this makes a normative approach even more important: to echo Nancy Fraser (1981, p. 286) once again, not "all forms of power are normatively equivalent", nor are "any social practices as good as any other". We hope that this collection allows us to begin to work towards conceptions and grounded practices of tolerance and conviviality that are, in some ways at least, 'better'.

Notes

1 Both authors contributed to this work in equal fashion. We would like to thank Aakash Singh Rathore for his valuable comments on the structuring of our argument.
2 For an account of practice-dependent political theory, see Sangiovanni (2008).
3 For a discussion of 'institutional practice-dependence' as opposed to a political realist version of practice dependence, see Gentile (2018).

References

Amin, A. (2004). Regions unbound: Towards a new politics of place. *Geografiska Annaler B*, 86(1), pp. 33–44.

Amin, A. (2006). The good city. *Urban Studies*, 43, pp. 1009–1023.

Back, L. and Sinha, S. (2016). Multicultural Conviviality in the Midst of Racism's Ruins. *Journal of Intercultural Studies*, 37(5), pp. 517–532.

Balint, P. (2017). *Respecting toleration: Traditional liberalism and contemporary diversity*. Oxford: Oxford University Press.

Barnett, C. (2012). Situating the geographies of injustice in democratic theory. *Geoforum*, 43, pp. 677–686.

Barnett, C. (2017). *The priority of injustice: Locating democracy in critical theory*. Athens, GA: University of Georgia Press.

Barnett, C. (2018). Geography and the priority of injustice. *Annals of the American Association of Geographers*, 108(2), pp. 317–326.

Bialasiewicz, L. (2006). 'The death of the west': Samuel Huntington, Oriana Fallaci and a new 'moral' geopolitics of births and bodies. *Geopolitics*, 11, pp. 701–724.

Bialasiewicz, L. (2017). That which is not a mosque: Disturbing place at the 2015 Venice biennale. *City*, 21(2/3), pp. 367–387.

Bracke, S. (2011). Subjects of debate: Secular and sexual exceptionalism and Muslim women in the Netherlands. *Feminist Review*, 98, pp. 28–46.

Bracke, S. (2012). From 'saving women' to 'saving gays': Rescue narratives and their dis/continuities. *European Journal of Women's Studies*, 19(2), pp. 237–252.

Brown, W. (2006). *Regulating aversion: Tolerance in the age of identity and empire*. Princeton, NJ: Princeton University Press.

Brown, W. and Forst, R. (2014). *The power of tolerance: A debate* (Di Blasi, L. and C. Holzhey, eds.). New York: Columbia University Press.

de Leeuw, M. and van Wichelen, S. (2012). Civilizing migrants: Integration, culture and citizenship. *European Journal of Cultural Studies*, 15(2), pp. 195–210.

Dittmer, J. (2007). Intervention: Religious geopolitics. *Political Geography*, 26, pp. 737–739.

Duyvendak, J.W., Geschiere, P. and Tonkens, E., eds. (2016). *The culturalizaton of citizenship: Belonging and polarization in a globalizing world*. London: Palgrave Macmillan.

Eisgruber, C. and Sager, L. (2007). *Religious freedom and the constitution*. Cambridge, MA: Harvard University Press.

Fassin, E. (2010). National identities and transnational intimacies: Sexual democracy and the politics of immigration in Europe. *Public Culture*, 22(3), pp. 507–529.

Forst, R. (2013). *Toleration in conflict: Past and present*. Cambridge: Cambridge University Press.

Fraser, N. (1981). Foucault on modern power: Empirical insights and normative confusions. *Praxis International*, 3, pp. 272–287.

Galston, W.A. (2010). Realism in political theory. *European Journal of Political Theory*, 9(4), pp. 385–411.

Gentile, V. (2018). Modus vivendi liberalism, practice-dependence and political legitimacy. *Biblioteca della Libertà BDL*, LIII(222), pp. 1–25.

Gilbert, D., Dwyer, C., Ahmed, N., Cuch-Graces, L. and Hyacinth, N. (2018). The hidden geographies of religious creativity: Place-making and material culture in West London faith communities. *Cultural Geographies*, 26(1), pp. 23–41.

Gray, J. (2000). *Two faces of liberalism*. New York: The New Press.

Gregory, D., Johnston, R., Pratt, G. and Watts, M., eds. (2009). *The dictionary of human geography*. 5th ed. Malden, MA: Wiley-Blackwell.

Hafez, F. (2018). Street-level and government-level Islamophobia in the Visegrád Four countries. *Patterns of Prejudice*, 52(5), pp. 436–447.

Harvey, D. (1976/2009). *Social justice and the city*. Revised ed. Athens, GA: University of Georgia Press.

Herzfeld, M. (2016). *Cultural intimacy: Social poetics and the real life of states, societies and institutions*. 3rd ed. New York: Routledge.

Ivakhiv, A. (2006). Toward a geography of 'religion': Mapping the distribution of an unstable signifier. *Annals of the Association of American Geographers*, 96, pp. 169–175.

Jones, P. (2017). Religious exemptions and distributive justice. In: C. Laborde and A. Bardon, eds., *Religion in liberal political philosophy*. Oxford: Oxford University Press, pp. 163–176.

Kalmar, I. (2018). 'The battlefield is in Brussels': Islamophobia in the Visegrad four in its global context. *Patterns of Prejudice*, 52(5), pp. 406–419.

Kong, L., Olson, E. and Hopkins, P., eds. (2013). *Religion and place: Landscape, politics, piety*. New York: Springer.

Krzyzanowski, M. (2013). From anti-immigration and nationalist revisionism to Islamophobia: Continuities and shifts in recent discourses and patterns of political communication of the freedom party of Austria. In: R. Wodak, M. Khosravinik, and B. Mral, eds., *Right-wing populism in Europe: Politics and discourse*. London: Bloomsbury.

Laborde, C. (2017). *Liberalism's religion*. Cambridge, MA: Harvard University Press.

Laborde, C. and Bardon, A., eds. (2017). *Religion in liberal political philosophy*. Oxford: Oxford University Press.

Lewicki, A. (2017). The blind spots of liberal citizenship and integration policy. *Patterns of Prejudice*, 51(5), pp. 375–395.

Lewicki, A. (2018). Race, Islamophobia and the politics of citizenship in post-unification Germany. *Patterns of Prejudice*, 52(5), pp. 496–512.

Massey, D. (1994). *Space, place and gender*. Minneapolis, MN: University of Minnesota Press.

Massey, D. (2005). *For space*. London: Sage.

McCabe, D. (2010). *Modus vivendi liberalism: Theory and practice*. Cambridge: Cambridge University Press.

McCrea, R. (2014). Religion in the workplace: *Eweida and others v. United Kingdom*. *The Modern Law Review*, 77(2), pp. 277–307.

Mitchell, D. (2003). *The right to the city: Social justice and the fight for public space*. London: Guilford Press.

Popke, J. (2006). Geography and ethics: Everyday mediations through care and consumption. *Progress in Human Geography*, 30(4), pp. 504–512.

Rawls, J. (1996). *Political liberalism*. New York: Columbia University Press.

Rheindorf, M. and Wodak, R. (2018). Borders, fences and limits - protecting Austria from refugees: Metadiscursive negotiation of meaning in the current refugee crisis. *Journal of Immigrant and Refugee Studies*, 16(1–2), pp. 15–38.

Rossi, E. and Sleat, M. (2014). Realism in normative political theory. *Philosophy Compass*, 9, pp. 689–701.

Sangiovanni, A. (2008). Justice and the priority of politics to morality. *Journal of Political Philosophy*, 16(2), pp. 137–164.

Sayer, A. and Storper, M. (1997). Ethics unbound: For a normative turn in social theory. *Environment and Planning D: Society and Space*, 15, pp. 1–17.

Sayyid, S. and Vakil, A., eds. (2010). *Thinking through Islamophobia: Global perspectives*. London: Hurst.

Scharffs, B.G. (2010). The freedom of religion and belief jurisprudence of the European court of human rights: Legal, moral, political and religious perspectives. *Journal of Law and Religion*, 26(1), pp. 249–260.

Scott, J.W. (2018). Hungarian border politics as an anti-politics of the European union. *Geopolitics*. DOI: 10.1080/14650045.2018.1548438

Tse, J. (2014). Grounded theologies: 'Religion' and the 'secular' in human geography. *Progress in Human Geography*, 38(2), pp. 201–220.

Williams, B. (2005). *In the beginning was the deed: Realism and politics in political argument*. Princeton, NJ: Princeton University Press.

Wodak, R. (2015). *Politics of fear: What right-wing populist discourses mean*. London: Sage.

Wodak, R., KhosraviNik, M. and Mral, B., eds. (2013). *Right-wing populism in Europe: Politics and discourse*. London: Bloomsbury.

Zucca, L. (2012). *A secular Europe: Law and religion in the European constitutional landscape*. Oxford: Oxford University Press.

Part I

Negotiating freedom and religion

Tolerance, neutrality, conviviality

1   The scope of religious freedom in Europe

Tolerance, democratic equality, and political autonomy[1]

Valentina Gentile

In October 2006, Nadia Eweida, a British citizen employed by British Airways, was placed on unpaid leave when she refused to conform to the company's uniform policy by openly wearing a necklace with a crucifix. Eweida's appeal to the British courts was rejected on the grounds that she had been offered another job that did not involve direct public contact with the company's customers and where she would have been permitted to wear the necklace; a position she had refused.[2] A couple of years earlier, Lillian Ladele, a registrar for marriages, births, and deaths for Islington Borough Council in London, was subjected to disciplinary action by her employer for not conforming to the Council's 'Dignity for All' equality and diversity policy after refusing to officiate at same-sex civil partnership ceremonies, citing her Christian faith. The British courts ultimately rejected Ladele's appeal alleging direct discrimination on the basis of religion or belief and harassment, ruling that the Borough of Islington's treatment of Ladele was a proportional means of achieving the legitimate aim of providing the registrar service on a non-discriminatory basis.[3] In 2000, Christiane Ebrahimian lost her job at the psychiatric unit of a public hospital in Paris when she refused to remove her Islamic headscarf at work. The French courts rejected Ebrahimian's complaints that the hospital director's decision infringed on her religious freedom by holding that it was directed to preserve the principles of the secularism and neutrality of public services.[4]

Religious freedom is widely claimed to be a fundamental principle of liberal democracy, yet the relation between religious freedom, tolerance, and democratic equality is a matter of ongoing negotiation. Citizens' requests to have special protections recognised on religious grounds in places of work, contestations regarding the appropriate constrains on religious practices and identities in schools, hospitals, and other public environments, and controversies about the admissibility of forms of discrimination (e.g. LGBTQ discrimination) on religious grounds abound in Europe and have given rise to several questions within and beyond national boundaries. Such controversies,[5] however, show that a liberal[6] understanding of religious freedom cannot be isolated from the ways in which the conception of tolerance, and the value of equality required by it, are justified to citizens. This chapter seeks to clarify the connections between two central tenets of liberal theory: first, a robust defence of citizens' freedom to pursue their religious commitments, and, second, an institutional understanding of tolerance and "democratic equality"[7] aimed at securing the stability 'for the right reasons' of liberal society. While liberal theory suggests that these two tenets are compatible, there is considerable tension between them.

In this chapter, I argue that the two central tenets of liberal democracy can work in harmony on the basis of a fundamental commitment to securing citizens' political autonomy and the political value of integrity connected to it. Political autonomy is a particular form of political freedom (see also Weithman, 2016, p. 168) which secures "the legal independence and assured political integrity of citizens and their sharing with other citizens equally in the exercise of political power" (Rawls, 1996, p. xlii). Liberal tolerance therefore ensures that all citizens will enjoy political autonomy while being free to decide about the weight to give to their moral commitments in light of their comprehensive views (1996, p. xliii). When citizens are politically autonomous, they recognise their role as self-legislators and, thus, their public and private – and therefore even religious – normative commitments merge. Certain religious integrity claims, however, might seem to endanger the allegiance of citizens of faith to the shared view of justice jeopardising their political autonomy: in these cases, liberal tolerance requires us to do what is most appropriate in a specific given situation. Thus, how free and equal citizens realise the political value of autonomy might reflect these different circumstances. In this chapter, I argue that the political value of autonomy does not merely permit but, rather, requires a degree of accommodation of citizens' religious claims in given situations. What matters is that such forms of protection reinforce citizens' allegiance to a shared conception of political justice.

Thus, the appeal to public justification does not necessarily entail the rejection of religious practices or certain forms of conscientious objection. Rather, some exemptions from ordinary laws on religious grounds should be taken as justifiable from a liberal viewpoint. This approach suggests that even if a law or regulation is neutral in its aim and, therefore, fully legitimate from the point of view of public reason, it might nonetheless have substantial burdening effects on citizens of faith. In these cases, all things considered, some forms of exemption should be justified as far as they do not undermine but rather are supportive of a political understanding of citizens as free and equal.

The first section provides an explanation of the nexus between tolerance, public justification, and religious protections. Following Schwartzman (2012, 2017, p. 17), this section shows that liberal egalitarian theories might be symmetric or asymmetric in the treatment of two important aspects at stake, namely public justification and religious accommodation. The second section clarifies the contours of the debate on liberal egalitarianism and the scope of religious freedom. The symmetric solution preferred by liberal egalitarians, especially Eisgruber and Sager's (2007) 'equal liberty', faces the problem of disregarding what is really at stake with religious exemptions. Building on Jones' (2017) criticism, this section argues that to determine the scope of religious freedom is a problem of non-comparative justice, one that guarantees the (equal) access to a constitutional right.

Yet, partially in contrast with Jones' view, the third section clarifies that this is still a problem of egalitarian justice: it aims to determine what is the most appropriate view of citizens' equality which helps to single out legitimate claims of religious integrity. The fourth section argues that Rawls' political value of autonomy and the related notion of democratic equality require a significant degree of accommodation of citizens' religious claims. In conclusion, the last section discusses the notion of political integrity as implied in the liberal commitment to political autonomy. It considers some recent attempts to combine liberal egalitarianism with a robust defence of personal integrity (especially Laborde, 2017). This section shows that there is no need to amend Rawls' public justification to defend a liberal commitment to citizens' religious integrity. The political value of integrity is meant to determine the appropriate constraints to religious freedom in contemporary democracies.

The political value of tolerance, public justification and religious protections

Contemporary liberalism and, most notably, Rawls' theory asks democratic institutions to reflect a conception of political authority which is based on a sophisticated view of tolerance (Rawls, 1996). Rawls' 'overlapping consensus' differs both from a merely pragmatic defence of coexistence understood as a *modus vivendi* (1996, p. xxxix–xli), and from independent (metaphysical/transcendental) ethical frameworks such as Christianity or Kantianism. The idea of an 'overlapping consensus' provides a common-ground view of tolerance that is *purely political* (Taylor, 1998, p. 37). The full and equal freedom of conscience and thought is one of its pillars (Rawls, 1996, p. xxvi). The protection of such basic freedoms is essential to ensure citizens' political autonomy and integrity and, therefore, their allegiance to just, liberal institutions. Key to Rawls' project, however, are the ideas of reasonableness and public reason: the plurality of reasonable religious or non-religious worldviews do not undermine the stability of the doctrine of democratic tolerance as far as the conception of justice is committed to public justification.

An important distinction emerges between citizens' actual religious views and practices, and the justification of the framework of rights and freedoms that guarantees their free exercise. Although citizens' disagreement about religion and particularly the ways that religious commitments and practices should be performed might be acute, the disagreement about the interpretation of the framework of freedoms and rights should nevertheless be committed to public justification. In Quong's words, the latter can be called a "justificatory disagreement", which occurs "when participants do share premises that serve as a mutually acceptable standard of justification, but they nevertheless disagree about certain substantive conclusions" (Quong, 2005, p. 303). As long as the form of disagreement about the understanding of tolerance, and the basic freedoms and rights attached to it, is justificatory, the authority of a liberal democracy is stable 'for the right reasons' and therefore fulfils the liberal principle of legitimacy.

However, it would be misleading to believe that this view of tolerance envisages a purely procedural understanding of neutrality.[8] As I mentioned earlier, it presupposes a common ground that is substantive rather than procedural, insofar as it expresses those political values that are affirmed in an "overlapping consensus" (Rawls, 1996, pp. 191–192). In this sense, this view of neutrality might be better understood as "a neutrality of aim": liberal institutions must secure the equality of opportunities to pursue any 'permissible' worldview, either religious or non-religious (Rawls, 1996, p. 193). Democratic societies must take the necessary measures – those that are consistent with the principle of freedom of conscience and thought – to support the political value of tolerance by discouraging attempts to discriminate on religious grounds. This view of neutrality envisages a wide protection to religious citizens by recognising extended accommodation of religious practices and ways of living in the private sphere, although the liberal principle of legitimacy requires that the justificatory structure of such freedom ought to be independent from any comprehensive (religious or non-religious) doctrine. This view should be distinguished from a "neutrality of effects or influence" (Rawls, 1996, p. 193), whereby it would be impossible to pretend that liberal institutions would have no effect on the stability of comprehensive doctrines of any kind. Liberal citizens should be able to revise their beliefs and commitments, but this inevitably endangers the stability of their comprehensive views. As long as liberal institutions' perverse effect does not single out a specific set of beliefs – say religions – then liberal institutions' neutrality of aim is not at risk. Yet if the neutrality of 'effects' is not plausible, there might be still doubts regarding what to do with those citizens who are particularly affected by certain laws or statutes that are neutral in aim.[9]

In the literature on liberal egalitarianism, the two dimensions of the value of tolerance give rise to two different debates. One is focussed on the meaning and scope of religious freedom. It asks whether liberalism should justify special protections to religious citizens and, more precisely, on what grounds a liberal democracy might be required to grant exemptions from generally applicable laws that are especially onerous for religious citizens. The issue here is to understand whether and how liberal democracies should accommodate citizens' religious claims in circumstances where the divide between public and private is somehow less visible (Galeotti, 2010). This situation occurs quite often in pluralistic societies. Consider, for example, the exemption of Sikhs from the law requiring motorcycle riders to wear a helmet in the United Kingdom, or the recognition by law of provisions aimed at protecting citizens from being compelled to perform certain actions, such as a doctor refusing to give an abortion[10] – i.e. protections from actions that are in conflict with one's deep convictions. The second debate regarding the value of tolerance refers to the public justification of freedoms and rights, including religious liberty, one that should ensure citizens' equality.

These two debates, however, are linked to each other in ways that are often neglected in the literature (though with some exceptions: see especially Kutoroff, 2015; Schwartzman, 2012, 2017 and Laborde, 2016, 2017) as far as they introduce what is said to be a paradox inherent to liberalism (Kutoroff, 2015, p. 248).

On the one hand, the political value of tolerance expressed in this view of neutrality implies that religions and non-religious worldviews are equally excluded from the point of view of public justification. In this first sense, liberalism does not accord any special status to religious doctrines in setting out those constraints to public deliberation which ensure citizens' political autonomy and equality. On the other hand, however, this same value seems to require states to grant special recognition to religious beliefs, practices, and identities. Yet, if religion is not special for tolerance in the first instance, why should it be in the second?

One way to begin considering this issue might be to ascertain whether religion is special for liberalism. A considerable amount of literature has been devoted to the alleged special status of religion[11] and whether liberalism should afford a special treatment to religions. This is a thorny issue, which risks displacing a first-order dispute regarding the content of political justice onto a second-order one regarding whether it is possible to single out religion on a morally relevant ground (see, for example, Brownlee, 2017). It is not the contention of this chapter to engage in this discussion. Rawls' liberalism is not a secular doctrine and its appeal to public justification does not single out religions from other non-religious comprehensive views (Rawls, 1996, p. 194 n24, 1997, pp. 771, 775, 779–780). The political conception of tolerance treats all worldviews, including non-religious ones, equally in this respect. Yet tolerance requires the protection of citizens' capacity to embrace, pursue, and revise their own conceptions of the good. This is particularly relevant for citizens of faith who want to pursue and live in accordance with their own religious beliefs.

As a matter of fact, religious claims are often given special consideration in contemporary democracies. People view religion as a fundamental and distinctive human commitment. Consider, for example, the reaction of British society to the Eweida case. Public opinion strongly supported her decision to refuse to conceal the cross. Several newspapers publicly condemned British Airways. And only three months after the controversy, BA instituted a new policy that permitted the display of all religious and charity symbols. Not only is there empirical evidence to support the significance of religion in contemporary democratic life, but there are normative arguments as well. Liberalism is primarily concerned with embracing and nurturing the pluralism of worldviews and identities, most of which are religious in nature. It would be at odds with tolerance to suppress religious practices and ways of life. Yet the ideal of public justification is often said to be in contrast with the possibility of cultivating religious diversity within democratic societies (see for example Vallier, 2012; Eberle, 2002; Wolterstorff, 1997a, 1997b).

At the core of this dispute is the alleged difficulty in combining the two faces of the political value of tolerance, which is defined on one level in terms of certain *restrictions* imposed on religions, and on another in terms of *protections* deserved by citizens of faith (for a useful taxonomy of this debate, see Schwartzman, 2017, p. 15ff). Different strategies might be employed in response. One might hold that religions should be excluded as grounds for public justification, but that they also deserve special protections.[12] This view, however, arbitrarily discriminates between religions and non-religious worldviews on both levels (see on this Schwartzman, 2017, p. 22). Other strategies might separate the two levels so that religions and religious reasons either should be *restricted* in public justification or should warrant *privileged treatment* in the form of legal exemptions. Asymmetric solutions, though, are inconsistent in their treatment of religions and raise the same equality objection for the arbitrary discrimination between religious and non-religious views mentioned previously (see Schwartzman, 2017, pp. 19–21). A great deal of recent work has been directed in responding to this problem. Central to this literature is the attempt to reconcile the liberal principle of equality with religious freedom (Eisgruber and Sager, 2007; Tebbe, 2017; Laborde, 2017; Schwartzman, 2012, 2017).

Liberal egalitarianism and the *scope* of religious freedom

This paradox has given rise to an intense debate regarding the meaning of religious freedom for liberal egalitarianism (Eisgruber and Sager, 2007; Tebbe, 2017; Laborde, 2017). A dominant thesis in this literature argues that the recognition of citizens' religious claims should be understood as part of the ideal of equality. This view does not accord any special status to religious commitments from the point of view of the accommodation: what counts is that a law or regulation does not infringe the general principle of "equal liberty" (Eisgruber and Sager, 2007, p. 13).

This is for instance the position of Eisgruber and Sager (2007), who argued that the main concern motivating the Supreme Court's reasoning in *Sherbert v. Verner*[13] was that of affirming a broader principle of equality. Adell Sherbert, a Seventh Day Adventist who lost her job because she refused to work on Saturdays for religious reasons, was denied access to unemployment benefits by the board of South Carolina, the state in which she lived. In a historic decision, the Supreme Court held that this denial violated the Free Exercise Clause. Sherbert's case has been often associated with an extended constitutional doctrine of religious freedom, one that affirms a robust defence of the Free Exercise Clause against the disestablishment of the state. As Eisgruber and Sager emphasise, the Court acknowledged that Sherbert and other religious citizens for whom Saturday is the Sabbath were victims of injustice, namely a radical unequal treatment compared to other employees of different religious faiths who were prevented from working on Sundays (Eisgruber and Sager, 2007, pp. 14–15). According to the two scholars, "equality was what was really at stake in Adell Sherbert's case", and it is because of equality that courts sometimes give special recognition to religious claims (2007, p. 15). In this way, Eisgruber and Sager circumvent the thorny dispute about the alleged special status of religion by providing a solution to the problem of reconciling the two dimensions of liberal tolerance that is symmetric: religion is not special for public justification so that non-public (religious) reasons might be invoked to justify political decisions and laws (Eisgruber and Sager, 2007, pp. 48–50), and neither is it special for accommodations (see also Schwartzman, 2017, p. 18).

Recently, Jones (2017) has incisively raised doubts regarding the emphasis placed on such an egalitarian interpretation of religious freedom. For Jones, if it is uncontroversial that nobody should be discriminated against on the basis of his or her religion, this view does not provide a satisfactory answer to the core question of religious exemptions, which refers to the scope of religious freedom in contemporary democracies. Jones claims that the issue does not lie in equality and, especially, in distributive justice; rather it refers to the scope of what he calls the "distribuend". He distinguishes between two kinds of "distribuend": a religious good might be defined as the "opportunity to embrace, pursue and live according to religion" (Jones, 2017, p. 164), whereas a non-religious good refers to the opportunity to enjoy goods that otherwise would be unavailable to individuals belonging to a certain faith. Different citizens' claims fall into these two categories. For example, the exemption of organised religions from discrimination law on grounds of gender in hiring their ministers is consistent with the aim to preserve their religious freedom and falls under the rubric of 'human rights law';[14] conversely, special provisions exempting Sikh people from wearing a helmet when they ride a motorcycle should be understood as a matter of 'non-discrimination law'.[15]

Cases like that of Sherbert illustrate two different kinds of wrong: Sherbert was a victim of a non-comparative injustice in being denied her constitutional right to religious freedom and of a distributive (comparative) injustice in being allowed less religious freedom than others.[16] However, what counts is to secure to Sherbert the free exercise of this constitutional right. This view helps to identify some important aspects involved in the Eweida case. Here, the ECHR forcefully argued that freedom of conscience, thought, and religion is a fundamental pillar of a democratic society. The judges also clarified that the religious dimension

> is one of the most vital elements that go to make up the identity of believers and their conception of life, but it is also a precious asset for atheists, agnostics, sceptics and the unconcerned. The pluralism indissociable from a democratic society, which has been dearly won over the centuries, depends on it.[17]

On this basis, the Court refused to hold that the possibility of Eweida taking the alternative role that was offered to her would negate the interference with that right. In that case, the Court explicitly stressed that "the better approach would be to weigh [the possibility of changing job] in the overall balance when considering whether or not the restriction [of religious freedom] was proportionate"[18] (on this point, see also McCrea, 2014, p. 279). For the Court, a proportionality test should give priority to the protection of religious freedom "as long as religious practices do not detrimentally affect service provision or unduly affect an employer".[19] The access to a religious good, to use Jones' expression, should take priority over other interests that, albeit legitimate, do not fall under the limits of Article 9(2) of the Convention.[20]

Jones suggests that comparative justice and, therefore, equality is not the issue in such cases. The *scope* of the distribuend rather than the distribution is what counts: the problem is how to weigh the different interests at stake. His thesis can be summarised as follows. Liberal egalitarianism (and especially Kymlicka, 1996; Quong, 2006; Eisgruber and Sager, 2007) defends religious exemptions as a corollary of the anti-discrimination law, and in these cases, they argue, egalitarian justice requires that the distribution of religious goods be allocated on a comparative basis so that equality of opportunity is secured. Yet religious freedom, to which all citizens are equally entitled, is a problem of non-comparative justice: if the state violates citizens' religious freedom, then we don't need to see whether the same violation has been perpetrated upon others to conclude that it is unjust.

The point is well taken. Religious exemptions should be justified on non-comparative grounds when the issue is the opportunity to access a religious good. What I find problematic, however, is the assumption that since religious exemptions are a matter of human rights law, the justification of exemptions should be taken separately from concerns about equality. As Jones (2017) concedes, in several cases it is difficult to distinguish clearly the scope of the distribuends. When a person is disadvantaged because of his or her religious view this translates into a violation of human rights law and of anti-discrimination law, but this should not make the distinction between the two distribuends less significant, Jones argues (2017, p. 166). One might think that, as the Court clarified in the Eweida case,[21] once the violation of human rights law is ascertained, in the first place, measures will be taken to avoid the discriminatory intent. So far so good. But there are cases where the free exercise might be invoked in contrast with other citizens' rights and freedoms. Here, the distinction between the two distribuends does not help to provide an answer to the problem raised by religious exemptions. Consider a situation in which a specific law or regulation poses a serious burden on a person's acting in accordance with his or her religious view, but the removal of such an obstacle might impact on other citizens' access to certain rights that should be secured on equal basis to all citizens. When this happens, a conflict arises between the free exercise that should be protected under human rights law, and egalitarian justice which requires that the state avoid discriminating on any ground whatsoever.

In some cases, the free exercise might demand forms of discriminatory anti-discrimination. However, even when forms of discrimination are necessary to access religious freedom, the justification of such exemptions should be extremely qualified. The so-called ministerial exception is useful to illustrate the point. For Jones, the exemption of organised religions from anti-discrimination law is an example of the protection of a 'religious good' that otherwise would be lost (Jones, 2017, p. 171). Yet, as Laborde (2017, p. 193) has argued, it would be too permissive to grant a full immunity to such religious groups; in her view, it should suffice to grant them deferential treatment,[22] namely protections accorded on the grounds of specific coherence and competence interests. Nevertheless, "they do not have the prerogative to determine their own sphere of autonomy" (Laborde, 2017, p. 196) – this is a matter that should be adjudicated by state authorities. Laborde's disaggregation theory (2017, p. 160ff) suggests that the problem of the 'ministerial exception' should be framed within the liberal value of freedom of associations instead. Jones could still object that such a move brings back a 'comparative' view of justice and therefore misses the point of the ministerial exception. Yet I believe that the scope of the ministerial exception is itself determined by a justificatory process that puts at its core a view of political autonomy, one that should ensure citizens' political and social equality.

In addition, citizens' claims to be exempted from anti-discrimination law on religious grounds should not be always permissible. In the case of Ladele, the applicant asked that a person with her religious belief be exempted from a policy which requires employees not to discriminate on the grounds of sexual orientation. Here, the proportionality test led the domestic courts to emphasise that the restriction of Ladele's religious belief had the legitimate aim of "protecting the rights of others which are also protected under the Convention".[23] In this case, considerations regarding the equality of citizens importantly informed the decision. In ruling against Ladele, the ECHR judges placed a significant emphasis on the 'margin of appreciation' and the ways in which democratic equality is endorsed and protected within European states. The judges acknowledged the serious burdens imposed on Ladele by the application of the law; but nevertheless they granted to British authorities a margin of appreciation when it came to striking a balance between competing fundamental rights: the right of Ladele not to be discriminated against on the grounds of religion versus the right of other citizens not to be discriminated against on the basis of sexual orientation (McCrea, 2014, p. 282). Thus, the securing of the equality of access to a non-religious good, namely civil partnership for same-sex couples, was considered superior to Ladele's actual loss of religious liberty.

Comparative justice, distributive justice, and citizens' equality

At the core of this dispute are different understandings of equality and, therefore, of comparative justice. Jones clarifies that he does not share Barry's (2001) aversion to religious exemptions; nonetheless, he agrees with him that religious exemption should not be a matter of distributive justice (Jones, 2017, p. 164). Thus, he employs a working definition of comparative justice that introduces a symmetry between comparative and distributive justice.[24] Yet I take Barry's argument for resisting religious exemptions as being strongly motivated by his account of equality, one that reflects a specific view of public justification. Interestingly, Quong noticed that the difference between Barry's view and his own (distributive) view regarding religious exemptions lies in their diverse understandings of public reason's scope. Quong argues that "Barry believes that public justifications only exist over a narrow range of issues, what Rawls calls the constitutional essentials and matters of basic justice, whereas as I argue that public reason can extend beyond this narrow range to address many other political issues" (2006, p. 60). Quong's defence of certain religious exemption is committed to what he calls a 'broad view' of public justification.

A different view of equality might suggest that treating people fairly requires the application of a differential treatment. Here, the case for exemptions is justified when a law or regulation imposes special burdens on some individuals or groups. Kymlicka's (1996) defence of certain religious accommodations has standardly been read in this way. Some religious exemptions are required by egalitarian justice to protect minorities who experience injustice because existing laws and policies reflect a distinctive culture, that of the majority. Yet, to overcome Barry's pragmatic account, this view must entail two further features: first, majorities have legitimate reasons to accommodate their own religious practices in crafting their laws and rules;[25] second, citizens have an intrinsic right to a particular treatment regardless of how other citizens are treated. Kymlicka treats religions as a sub-category of broader cultures.[26] He suggests that cultures are instrumentally significant to realise autonomy. Contemporary liberal-nationalist projects have legitimately sought to secure solidarity and political legitimacy also through the promotion of what he calls a shared 'societal culture'[27] (Kymlicka, 1996, p. 76ff). However, by giving priority to one cultural option, contemporary democracies are likely to undermine the autonomy of some citizens who belong to other cultural groups. Hence, standard liberal citizenship does not suffice to fulfil the demand of equality within diverse societies (Kymlicka, 1996, p. 45). Kymlicka's defence of exemptions also involves another aspect. Significant emphasis is placed on the idea that a baseline notion of equality requires the removal of unequal and unfair effects of brute bad luck on individuals' prospects. Quong rightly notices that "Kymlicka proposes luck egalitarian reasons for compensating those unlucky enough for having been born into cultures that are disadvantaged" (2006, p. 55). Yet this account of equality is independent of comparative concerns of justice: if certain groups are born in disadvantaged minorities, say indigenous or national minorities, they should be compensated.

We are thus presented with two different ways of considering the relation between religious exemptions and equality. For Barry, equality requires an equal treatment, yet some exemptions might be tolerated for pragmatic reasons. The argument assumes that either the case for the exemption is so strong as to rule out the general law, because it is openly discriminatory, or the case for the law is strong enough in ensuring equality of treatment to rule out the exemption. With reference to the Sikhs' exemption from wearing a helmet, Barry argues that "retaining the law but rescinding the exemption would have good direct effects but would probably have bad effects on the relations among religious groups and would give an encouraging signal to racists who peddle nostalgic dreams of cultural homogeneity" (2001, p. 50). Kymlicka instead justifies religious exemptions only in cases where individuals' circumstances are clearly constrained by the condition of unequal and unfair effects of bad brute luck. Yet such a luck-egalitarian view does not help to adjudicate cases in which citizens legitimately claim the protection of their religious freedom, but religion is constructed as a matter of choice: here, religious claims should be considered as an expensive taste in the standard luck-egalitarian debate (on this point see also Quong, 2006, p. 56).

I agree with Jones that Kymlicka's view does not help to grasp what is at stake when we try to determine the scope of religious liberty. Yet I don't believe the problem lies in the fact that he considers religious exemptions as a matter of equality. It is how citizens' equality and autonomy are constructed in this approach that I find problematic. Kymlicka believes that some preferences, including religions and cultures, might be externally constrained. Thus, state neutrality should be amended to guarantee that 'unfortunate' religious minorities have access to their own 'context of choice' (Kymlicka, 1996, p. 83). Yet the opportunity to embrace, pursue, and live according to a 'permissible' religious or non-religious form of perfection is a value that should be ensured by liberal institutions independently of whether religious aspects are constructed as unchosen circumstances. This view was persuasively suggested by the ECHR's ruling on the Eweida case when the Court argued that the opportunity offered to Eweida to change her job was not enough to avoid the interference with her religious freedom.

Another sense of equality seems to be relevant here. As Anderson has argued, it is not morally plausible to maintain that justice requires individuals to be compensated for their disadvantages resulting from certain adverse circumstances but never for those arising from their own choices (Anderson, 1999; Scheffler, 2003, p. 18). A more promising account of justice, and therefore of equality, is one that reflects the eminently relational structure of liberal democracies. 'Democratic equality' (Anderson, 1999, p. 289) is a moral ideal that governs the relationships of democratic citizens (Scheffler, 2003, p. 22). Scheffler has persuasively argued that this form of equality introduces a moral, social, and political dimension (2003, pp. 22–23). From a moral point of view, equality requires that all people are recognised as having equal moral worth by virtue of their status as persons. As a social ideal, it entails a notion of society understood as a cooperative system where its members retain the same social standing. Finally, as a political ideal, equality defines those claims that political agents "are entitled to make one another by virtue of their status as citizens" (Scheffler, 2003, p. 22). From the point of view of democratic equality, concerns of distribution are important simply because certain distributive arrangements are inconsistent with that social and political value. Significant inequalities in the distribution would easily translate into inequalities of power and status that are incompatible with an ideal relation among citizens as free and equal (see Scheffler, 2003, p. 29).

Quong (2006, p. 58ff) also refers to this view of equality when he discusses the problem of religious exemptions. He departs from Rawls' neutrality of aim and proceeds by arguing that when a religious citizen is substantially burdened by a law or regulation, which affects the basic opportunities to which he is entitled *qua* citizen, a religious exemption is a requirement of justice (Quong, 2006, p. 58). However, distributive justice is not the direct focus of such a view of equality. The idea of democratic equality suggests that citizens should be given equal opportunities to pursue 'permissible' conceptions of the good within a framework that ensures reciprocity. The idea is that when certain demands or commitments, especially those we think of as 'doctrinal', are not accommodated by laws or statutes, this might have the perverse effect of curtailing citizens' political integrity and, more broadly, their capacity to act with political autonomy. This perverse effect is one that political liberalism should not allow. In the next section, I link the problem of religious exemptions to the political value of autonomy.

Democratic equality and citizens' political autonomy

Jones (2017) believes that where the purpose of a legal exemption is to secure religious freedom, the exemption typically concerns the scope of the distribuend rather than its distribution. In such cases, religious exemptions should be understood as a matter of non-comparative justice, which ensures all citizens (equal) access to a constitutional right. I think this is a good point. However, the interpretation of this constitutional right is also subjected to a shared form of public justification. Thus access to religious freedom is still a matter of equality: we must determine what the most appropriate view of citizens' equality is that would single out legitimate claims of religious integrity. Here my view diverges from other liberal egalitarian strategies, like that of Quong (2006), who derives from Rawls' neutrality of aim a different distributive answer to the problem of religious exemptions: one that ensures a fair equality of opportunity. In my view, a liberal neutrality of aim and its shared justificatory structure ensures that citizens are able to act with political autonomy. Political autonomy is meant to protect a special form of political freedom. As Weithman nicely puts it, this form of political freedom "is realised, not by a person in the whole life, but by citizens in the political life" (Weithman, 2016, p. 168). Political autonomy is the ground on which citizens' self-determination is realised. Yet collective self-determination is tied to citizens' capacity to regard laws and statuses as supported by reasons of the right kind.

Public reasons ensure that citizens are equally politically free in this important respect, so that a reconciliation between their public role as citizens and their private interests and normative commitments is, at least, conceivable. Democratic equality, in turn, requires that citizens recognise their reciprocal responsibilities as derived from this shared understanding of political autonomy. Rawls argues that political autonomy implies "the legal independence and assured political integrity of citizens and their sharing with other citizens equally in the exercise of political power" (1996, p. xlii). For Rawls, political justice affirms a value of political autonomy for all citizens, while at the same time leaving to citizens to decide the appropriate weight to give to their moral understanding of autonomy (and, therefore, integrity) in light of their comprehensive views (1996, pp. xliii–xliv).

The value of political autonomy implies, on the one hand, that the justification of the framework of liberal rights and freedoms should be public so that non-public, and therefore also religious, reasons cannot serve as a basis for justifying the shared conception of justice; and, on the other, that religious views and practices significantly inform citizens' integrity, and therefore should be taken under great consideration when defining the meaning and scope of such freedoms, and of religious freedom in particular. One might wonder whether from such a framework it is possible to derive a practice of tolerance that effectively includes individuals' ways of life and religious practices. Williams, for example, argues that "it is very difficult both to claim that the value of autonomy is the foundation of the liberal belief in toleration, and at the same time to hold, as Nagel and Rawls and other liberals hold, that liberalism is not just another sectarian doctrine" (Williams, 2005, p. 131). He seems to be unconvinced about the possibility of distinguishing a political value of autonomy from a more robust, Kantian-inspired, moral ideal. However, if it is true that the account of justice as fairness presented in *A Theory of Justice* (1971) was fully justified by a (partially) comprehensive account of autonomy of the kind of Kant or Mill, the 'new' political framework of justice presented by Rawls in *Political Liberalism* (1996, pp. xli n8, xliii–xliv, 1997, p. 778) insists on the distinction between the political value of autonomy from other moral views of autonomy, such as Mill's idea of individuality or Kant's doctrine of autonomy. Political autonomy is constrained by reciprocity and allows citizens who hold different religious worldviews to see the compatibility, when not the congruence, between the *right* (and, therefore, the legitimacy of the shared conception of political authority) and the *good* of their different (non-public) normative commitments.

Therefore, the political value of autonomy not merely permits but rather requires a degree of accommodation of citizens' religious claims. What matters is that these protections reinforce of citizens' allegiance to the shared conception of justice. This, for example, is what Rawls says about the Quakers' objection to war in his 'Reply to Habermas' (Rawls, 2005, p. 393ff). Here, the Quakers' conscientious refusal to wage war might not be inconsistent with their endorsement of liberal institutions as legitimate. Rawls argues that "this case shows how political values can be overriding in upholding the constitutional system itself, even if particular reasonable statutes and decisions might be rejected" (2005, p. 394). Similar reasoning is provided with reference to the case of abortion. In an oft-quoted passage (Rawls, 1997, pp. 178–179), Rawls argues that citizens of faith who strongly reject permissive laws on abortion might nonetheless recognise the legitimacy of the law in its justificatory premises. Here, Rawls clarifies that the existence of such laws does not prevent citizens of faith from exercising forms of conscientious refusal or, in his words, 'they need not themselves exercise the right to abortion' (1997, p. 179). One might object that this way of reasoning raises problems of stability insofar as it is difficult to see how citizens of faith who might be asked to support financially, via taxes, the medical provision of abortion might accept it as a legitimate demand (Horton, 2003, p. 19). In response, it should be enough to note that permissive laws on abortion have been enacted in several European countries and yet the possibility of refusing to perform an abortion should not (and does not, in fact) affect citizens' general willingness to support, even financially via taxes, the implementation of such laws.[28]

Returning to the Eweida case, the British tribunals failed to weigh adequately the different interests at stake. The Employment Tribunal denied that the case displayed indirect discrimination as the visible wearing of a cross was not a mandatory requirement of the Christian faith but rather Eweida's personal choice. The Employment Appeal Tribunal added that the concept of indirect discrimination implies alleged unfairness towards a defined group, in this case the group of Christians within BA, and that the applicant had not established evidence of group disadvantage. A different view was, nonetheless, maintained by the Court of Appeal, according to which the previous tribunal failed to provide an account of indirect discrimination which did take into consideration the evidence of disadvantage to a single individual. Yet this court rejected Eweida's claim of indirect discrimination on the basis that she was offered to change job without loss of pay, though she rejected it. Thus, the Court of Appeal maintained that there can be no charge of interference with the right to manifest a religious belief where a person chooses voluntarily to accept a job which does not accommodate that practice when there are other means open to the person to practise or observe his or her religion. According to the ECHR, however, the British tribunals failed to find a fair balance between Eweida's fundamental interest to manifest her faith openly[29] and the legitimate, albeit less relevant, interest of British Airways to promote a certain corporate image.

Political integrity and the limits of religious freedom

In this chapter, I contended that the political value of tolerance not merely permits but rather requires some degree of accommodation of religious claims and that this freedom should be guaranteed on a non-comparative basis. At the same time, I argued that this is still a matter of egalitarian justice and, precisely, it reflects a commitment to public justification. Rawls' appeal to public reason ensures that citizens act with political autonomy, that is "the legal independence and assured political integrity of citizens and their sharing with other citizens equally in the exercise of political power" (Rawls, 1996, p. xlii). Thus, the problem with religious exemptions reflects a commitment to 'political integrity' implicit in the notion of political autonomy. What then is the meaning of 'political integrity' and how is this related to citizens acting autonomously in political terms?

The appeal to integrity has been invoked to criticise the constraints of liberal public reason on citizens of faith (Vallier, 2012). A well-known example of the integrity objection was formulated by Wolterstorff who argued that by privatising religious convictions Rawls' liberalism disregarded a fundamental aspect of religions, namely their "wholeness, integrity, integration, in [religious citizens'] lives" (1997b, p. 105). Recently, liberal egalitarian scholars have employed this argument based on the value of integrity to justify some religious exemptions (see especially Laborde, 2017; Seglow, 2017; Vallier, 2016; Bou-Habib, 2006). Laborde, in particular, combines a weaker understanding of public reason, what she calls public reason *stricto sensu* – one that emphasises the principle of 'public accessibility' (Laborde, 2017, p. 120ff) – with an egalitarian defence of religious exemptions precisely on the grounds of personal integrity. Following Paul Bou-Habib (2006), Laborde (2018, p. 14) believes that integrity has the advantage of appealing to a good that is clearly identifiable by non-religious citizens as well. Personal integrity might be said to be valuable for a number of reasons, including autonomy, identity, and self-respect among the others (see Lenta, 2016; Seglow,

2017). Thus the protection of certain 'integrity-related claims' should not be problematic for liberal egalitarianism.

In contrast to several critics of Rawls' public reason (Eberle, 2002; Wolterstorff, 1997a, 1997b), I agree with Laborde that Rawls' liberalism need not be inconsistent with citizens' religious integrity. What I find problematic in Laborde's view is the move from Rawls' requirement of citizens' endorsement – shareability – to a weaker form of 'citizens' understanding' via the accessibility principle (Laborde, 2017, p. 120ff). Laborde hopes that by amending the constraints of public reason the liberal legitimacy of democratic institutions is not undermined by the public recognition of some integrity-related claims as a matter of equality of treatment. Yet, in emphasising what I take to be a 'thick' epistemic desideratum, namely accessibility, Laborde's solution gets closer to the view of public justification envisaged by Audi (2014) – that is, an epistemic defence of secular reasons – rather than to Rawls's public justification. Laborde in fact provides a different kind of asymmetric solution to the problem discussed in this chapter: one that singles out what she calls a 'theocratic' dimension of religion from the purpose of public justification, but which does not grant the same special role to religion (Laborde, 2016, 2017). At this this second level, she seems to share Eisgruber and Sager's view that some exemptions, on the grounds of personal integrity, should be seen as a matter of equal liberty.

Yet, as Jones (2017) has argued, religious exemptions raise a problem of non-comparative justice: that of ensuring citizens' access to a constitutional right, namely religious freedom. From the point of view of liberalism, what counts is that the justification of this freedom is one that ensures citizens' allegiance to the fair scheme of social cooperation over time. Rawls's justificatory framework guarantees that a privileged form of stability is achieved and citizens act with political autonomy. The political value of autonomy, in turn, is capacious enough to value citizens' integrity. As the examples of Quakers and pro-life Catholics show, the ideal of liberal legitimacy is not always endangered by forms of conscientious refusal. Rather, liberalism requires us to do what is most appropriate in given situations: in this sense, how free and equal citizens realise the value of political autonomy might be situation sensitive. Importantly, however, integrity here is defined as a political value too – one which should be bounded by reasonableness.

Traces of this reasoning can be found in the ECHR's decision about the Eweida case. Here, the Court acknowledged that there was a violation of Eweida's rights in breach of Article 9 of the Convention. The Court favours an interpretation of individual-based integrity when it places important weight on "the value to an individual who has made religion a central tenet of his or her life to be able to communicate that belief to others".[30] Yet the political dimension of integrity introduces important constraints on the kind of justifiable claims for exemptions. By political, I mean that the claims of integrity should not trump principles such as reciprocity and democratic equality on which the conception of justice of a democratic society stands. This aspect informed the reasoning of the British judges in ruling against Ladele's claim for conscientious objection to registering same-sex civil unions. Here, to favour Ladele's religious objection over other citizens' right to access to such a good as the civil union on equal basis might have the effect of curtailing the view of neutrality of aim of liberal institutions.

A different way of interpreting political integrity seems to emerge from the ECHR reasoning in the case of Ebrahimian. Her wish to wear the headscarf at work was said to be incompatible with the principle of neutrality applied to public services. In the case of Ladele, a 'wide margin of appreciation' was employed by the Court to determine the boundaries between state and religion and the scope of religious freedom.[31] Yet significant objections precisely on the grounds of citizens' equality were raised from some of the members of the Court in the Ebrahimian decision. In his dissenting opinion, Judge De Gaetano argued that the French law on neutrality of public service should be understood as requiring neutrality of public officials with regard to "the *subjective* manifestation of one's religious belief and not the *objective* wearing of a particular piece of clothing or other symbol".[32] For the judge, requiring a public official to 'disclose' whether an item he or she wears is a manifestation of his or her religious view is contrary to the form of respect that the state owes each citizen towards his or her conscience. In her separate – partially dissenting – opinion, Judge O'Leary recalled that the Court's appeal to the states' margin of appreciation should not absolve domestic authorities from the obligation to concretely assess the proportionality of the measure taken "particularly when it is an issue of a blanket ban which interferes with the rights of an individual, while also affecting the employment opportunities of an entire collectivity".[33]

In conclusion, the proportionality test suggested by Judge O'Leary should be one that ensures that citizens' opportunity to live, embrace, and pursue their religious beliefs is consistent with a certain understanding of equality in relations with other citizens understood as free and equal. The principle of neutrality applied to public service should thus be understood in a very limited way to guarantee that no religious or non-religious worldview is endorsed by the state, while at the same time it should recognise for public servants the maximum freedom of conscience compatible with their role. Therefore, I agree with Judges O'Leary and De Gaetano who doubt that the emphasis on the *ostentatious* manifestation of one's religious beliefs should be seen as a good reason for restricting public servants' religious freedom.

Notes

1 For their stimulating comments, I am grateful above all to Luiza Bialasiewicz, Vittorio Bufacchi, Cristina Fasone, Megan Foster, Corrado Fumagalli, Anne Hewitt, Domenico Melidoro, and Aakash Singh Rathore. Special thanks are owed to Peter Jones for his extensive comments and valuable suggestions for the final draft of this chapter.
2 *Eweida and Others v. The UK* (2013) ECHR n.16.
3 *Eweida and Others v. The UK* (2013) ECHR n.29.
4 *Ebrahimian v. France* (2015) ECHR n. 16.
5 It should be noted that the three cases discussed in the chapter are different: Eweida was a private employee while, for example, Ebrahimian served as nurse in a public

hospital. These differences do not, nevertheless, affect the main argument suggested here. This chapter seeks to show that the kind of normative reasoning that determines the scope of religious freedom is connected to a process of public justification aimed at securing political autonomy of citizens and the related ideal of democratic equality.

6 In this chapter I employ the term 'liberal' or 'liberalism' to refer to Rawls' political liberalism (1996, 1997). This view is confronted by a family of liberal egalitarian theories (including Kymlicka, 1996; Barry, 2001; Quong, 2005, 2006; Eisgruber and Sager, 2007; Schwartzman, 2012; Audi, 2014; Laborde, 2017; Tebbe, 2017). I will refer to the latter as 'liberal egalitarianism'.

7 By democratic equality I mean an understanding of equality that is eminently relational, one presented by Anderson (1999) and Scheffler (2003), among others. In this account, 'democratic equality' is meant to contrast oppression, status hierarchies and privileges, and its aim is 'to create a community in which people stand in relations of equality to others' (Anderson, 1999, p. 289).

8 It should be noted that Rawls' use of the term neutrality – which he finds "unfortunate" (1996, p. 191) – in *Political Liberalism* is very limited. He refers to it only once (1996, pp. 191–196) to clarify how the priority of the 'right' relates to the ideas of the 'good' used in justice as fairness (1996, p. 173ff).

9 Samuel Freeman has recently emphasised such a potential "disruptive" character of liberal neutrality in his paper 'Democracy, Religion, and Public Reason' presented at the Global Issues in Ethics III: Religion and Democracy conference organised by the Australian Catholic University, Rome, March 2019.

10 Of the EU member states in which induced abortion is legal, invoking a conscientious objection is granted by law in 21 countries. For a detailed discussion of the CO in Europe see Heino et al. (2013).

11 For a reconstruction of this debate see especially Lund (2017), Cornelissen (2012, p. 85) and Ellis (2006, p. 219). See also Laborde (2016) and Laborde and Bardon (2017).

12 An example of a symmetric view of this sort is provided by Audi (2014).

13 *Sherbert v. Verner*, 374 U.S. 398 (1963).

14 Jones's aim here is to show that where the purpose of a legal exemption is to protect and secure religious freedom, the exemption will typically concern the scope of the distribuend rather than its distribution. I return to this point later in this section.

15 Jones believes that the case of Sikh motorcyclists is *like* non-discrimination law in the sense that the aim of the exemption is that of securing a non-religious good (namely, the option of riding a motorbike) rather than protecting a religious practice. However, if an exemption were not granted in this case, this would not translate into a form of discrimination against Sikhs' religious practices.

16 This case was discussed by Jones in the unpublished version of his paper, 'Religious Exemptions and Distributive Justice', presented at the RAPT conference in London in June 2015.

17 *Eweida and others v. the United Kingdom* 37 (2013), n. 79.

18 *Eweida and others v. the United Kingdom* 37 (2013), n. 83.

19 *Eweida and others v. the United Kingdom* 37 (2013), n. 78.

20 Art. 9(2) states what follows: "Freedom to manifest one's religion or beliefs shall be subject only to such limitations as are prescribed by law and are necessary in a democratic society in the interests of public safety, for the protection of public order, health or morals, or for the protection of the rights and freedoms of others" (Council of Europe, 1950).

21 *Eweida and others v. the United Kingdom* 37 (2013), n. 95. Here the judge argued that "In the light of this conclusion, [the Court] does not consider it necessary to examine separately the applicant's complaint under Article 14 taken in conjunction with Article 9".

22 A different view of deferential treatment has been recently employed by Paul Weithman. In his view, a "non-deferential treatment" should be accepted by citizens of ecclesial faith towards ecclesiastic hierarchies and clergy. This attitude better conforms to political liberalism as far as it is aimed at guaranteeing special protection of those "who are in vulnerable positions with security and to protect them from what are uncontroversial bads". 'Liberal Society and Deferential Treatment' paper presented at the Global Issues in Ethics III: Religion and Democracy conference organised by the Australian Catholic University, Rome, March 2019.

23 *Eweida and others v. the United Kingdom* 37 (2013), n. 104.

24 Though he concedes that this supposition is not completely correct, as in the case of desert theories that are both non-comparative and distributive (Jones, 2017, p. 164).

25 As White (2012, p. 8) has suggested, if the majority act illegitimately to craft their laws, we should not claim that justice requires the extension of accommodation, via exemptions, to other religious groups. This situation would rather require that the state should refrain from crafting such a law.

26 Kymlicka has been often criticised for neglecting the significance of religions: see for instance Modood (1998).

27 Kymlicka defines a "societal culture" as that culture "which provides its members with meaningful ways of life across the full range of human activities, including social, educational, religious, recreational, and economic life, encompassing both public and private spheres" (1996, p. 76).

28 The case of abortion is particularly controversial. In various European states, special provisions regarding conscientious objections for medical doctors and nurses are recognised by law. To grant a conscientious objection to medical doctors might have the perverse consequence of putting at risk the effective implementation of the law when the number of objectors is very high. However, this difficulty should not be troubling for political liberalism. Rawls believes that insofar as pro-life Catholics are committed to public justification, their refusal of abortion is not one that undermines the legitimacy of liberal democratic institutions. He hopes that once a permissive law on abortion is democratically enacted and protections are granted to the minority against abortion, so that pro-life Catholics can freely express their dissent in line with public reason, the law in question will be stable.

29 The Court reiterated the point that "this is a fundamental right: because a healthy democratic society needs to tolerate and sustain pluralism and diversity; but also because of the value to an individual who has made religion a central tenet of his or her life to be able to communicate that belief to others". *Eweida and others v. the United Kingdom* 37 (2013), n. 94.

30 *Eweida and others v. the United Kingdom* 37 (2013), n. 94.

31 *Ebrahimian v. France* (2015) ECHR n. 70.

32 *Ebrahimian v. France* (2015) Separate Opinions. Dissenting Opinion of Judge De Gaetano, p. 39.

33 *Ebrahimian v. France* (2015) Separate Opinions. Dissenting Opinion of Judge O'Leary, pp. 37–38.

References

Anderson, E. (1999). What is the point of equality? *Ethics*, 109(2), pp. 287–337.

Audi, R. (2014). *Democratic authority and the separation of church and state*. Oxford: Oxford University Press.

Barry, B. (2001). *Culture and equality*. Cambridge: Polity Press.

Bou-Habib, P. (2006). A theory of religious accommodation. *Journal of Applied Philosophy*, 23(1), pp. 109–126.

Brownlee, K. (2017). Is religious conviction special? In: C. Laborde and A. Bardon, eds., *Religion in liberal political philosophy*. Oxford: Oxford University Press, pp. 309–320.

Cornelissen, G. (2012). Belief-based exemptions: Are religious beliefs special? *Ratio Juris*, 25(1), pp. 85–109.

Council of Europe. (1950). European convention for the protection of human rights and fundamental freedoms (as amended by Protocols Nos. 11 and 14, 4 November 1950). *ETS 5*. Available at: https://www.refworld.org/docid/3ae6b3b04.html. [Accessed 20 Aug. 2019].

Eberle, C.J. (2002). *Religious conviction in liberal politics*. New York: Cambridge University Press.

Eisgruber, C.L. and Sager, L.G. (2007). *Religious freedom and the constitution*. Cambridge, MA: Harvard University Press.

Ellis, A. (2006). What is special about religion? *Law and Philosophy*, 25(2), pp. 219–241.

Galeotti, A.E. (2010). The place of conscientious objection in a liberal democracy. In: G. Calder and E. Ceva, eds., *Diversity in Europe: Dilemmas of differential treatment in theory and practice*. London: Routledge, pp. 17–31.

Heino, A., Gissler, M., Apter, D. and Fiala, C. (2013). Conscientious objection and induced abortion in Europe. *The European Journal of Contraception and Reproductive Health Care*, 18(4), pp. 231–233.

Horton, J. (2003). Rawls, public reason and the limits of liberal justification. *Contemporary Political Theory*, 2(1), pp. 5–23.

Jones, P. (2017). Religious exemptions and distributive justice. In: C. Laborde and A. Bardon, eds., *Religion in liberal political philosophy*. Oxford: Oxford University Press, pp. 163–176.

Kutoroff, A. (2015). First Amendment versus *Laïcité*: Religious exemptions, religious freedom, and public neutrality. *Cornell International Law Journal*, 48, pp. 247–278.

Kymlicka, W. (1996). *Multicultural citizenship: A liberal theory of minority rights*. Oxford: Oxford University Press.

Laborde, C. (2016). Conclusion: Is religion special? In: J.L. Cohen and C. Laborde, eds., *Religion, secularism and constitutional democracy*. New York: Columbia University Press, pp. 423–433.

Laborde, C. (2017). *Liberalism's religion*. Cambridge, MA: Harvard University Press.

Laborde, C. (2018). Three cheers for liberal modesty. *Critical Review of International Social and Political Philosophy*, pp. 1–17.

Laborde, C. and Bardon, A. (2017). Introduction. In: C. Laborde and A. Bardon, eds., *Religion in liberal political philosophy*. Oxford: Oxford University Press, pp. 1–11.

Lenta, P. (2016). Freedom of conscience and the value of personal integrity. *Ratio Juris*, 29(2), pp. 246–263.

Lund, C. (2017). Religion is special enough. *Virginia Law Review*, 103, pp. 481–524.

McCrea, R. (2014). Religion in the workplace: *Eweida and others v. United Kingdom*. *The Modern Law Review*, 77(2), pp. 277–307.

Modood, T. (1998). Anti-essentialism, multiculturalism and the "recognition" of religious groups. *Journal of Political Philosophy*, 6, pp. 378–399.

Quong, J. (2005). Disagreement, asymmetry, and liberal legitimacy. *Politics, Philosophy and Economics*, 4(3), pp. 301–330.

Quong, J. (2006). Cultural exemptions, expensive tastes, and equal opportunities. *Journal of Applied Philosophy*, 23(1), pp. 53–71.

Rawls, J. (1971). *A theory of justice*. Cambridge, MA: Harvard University Press.

Rawls, J. (1996). *Political liberalism*. New York: Columbia University Press.

Rawls, J. (1997). The idea of public reason revisited. *The University of Chicago Law Review*, 64(3), pp. 765–807.

Rawls, J. (2005). *Political liberalism*. Expanded ed. New York: Columbia University Press.

Scheffler, S. (2003). What is egalitarianism? *Philosophy and Public Affairs*, 31(1), pp. 5–39.

Schwartzman, M. (2012). What if religion is not special? *University of Chicago Law Review*, 79(4), pp. 1351–1428.

Schwartzman, M. (2017). Religion, equality, and anarchy. In: C. Laborde and A. Bardon, eds., *Religion in liberal political philosophy*. Oxford: Oxford University Press, pp. 15–30.

Seglow, J. (2017). Religious accommodation: Responsibility, integrity and self-respect. In: C. Laborde and A. Bardon, eds., *Religion in liberal political philosophy*. Oxford: Oxford University Press.

Taylor, C. (1998). Modes of secularism. In: R. Bhargava, ed., *Secularism and its critics*. New Delhi: Oxford University Press, pp. 31–70, 177–190.

Tebbe, N. (2017). *Religious freedom in an egalitarian age*. Cambridge, MA: Harvard University Press.

Vallier, K. (2012). Liberalism, religion and integrity. *Australasian Journal of Philosophy*, 90(1), pp. 149–165.

Vallier, K. (2016). The moral basis of religious exemptions. *Law and Philosophy*, 35, pp. 1–28.

Weithman, P. (2016). *Rawls, political liberalism and reasonable faith*. Cambridge: Cambridge University Press.

White, S. (2012). Religious exemptions: An egalitarian demand? *The Law and Ethics of Human Rights*, 6(1), pp. 97–118.

Williams, B. (2005). *In the beginning was the deed: Realism and politics in political argument*. Princeton, NJ: Princeton University Press.

Wolterstorff, N. (1997a). The role of religion in decision and discussion of political issues. In: R. Audi and N. Wolterstorff, eds., *Religion in the public square: The place of religious convictions in political debate*. New York: Rowman and Littlefield, pp. 67–120.

Wolterstorff, N. (1997b). Why we should reject what liberalism tells us about speaking and acting in public for religious reasons. In: P. Weithman, ed., *Religion and contemporary liberalism*. Notre Dame, IN: University of Notre Dame Press, pp. 162–181.

2   Neutrality, toleration, and religious diversity

Peter Balint

A generally tolerant state can have a range of relationships (respect, indifference, or forbearance tolerance) to the particular ways of life of its citizenry and still be accommodating of their difference. This is because the practice of toleration can be understood in at least two different ways. The first, where the tolerating agent must in some way object to the thing being tolerated; and the second one where an objection is not required. This second understanding makes sense of terms such as 'a tolerant society', where the toleration that takes place usually involves much more indifference than it does grudging acceptance.[1] When we say the state (or their agent) should be tolerant of difference, in this case religious difference, we are not, then, specifying what this might mean; a state may respect some differences (for example, the place of Anglicanism in the United Kingdom), forbear others (perhaps some of the practices of the Amish on the United States), and may be entirely indifferent to others (for example, the practices of Baptism or Bat Mitzvah). Being generally tolerant allows this range of response to specific religious differences. The question then is, is there a normative ideal? In this chapter, I want to argue that a particular kind of neutrality – what I will call 'active indifference' – should be the dominant form of liberal toleration, and that contrary to much of the multicultural literature, respect for difference should not be the ideal.

Challenging neutrality

Indifference, particularly (but not only) in multicultural circles, has had such a sustained critique over the past couple of decades that its failure to fairly deal with diversity is now taken as a received view. Indeed its failure to address issues of diversity is a crucial premise in the argument of the most dominant schools of multiculturalism and group-differentiated rights.[2] Their claim is that the contemporary liberal state inevitably privileges some ways of life, generally those of the majority of citizens, over others, usually those of minorities. The type of indifference that Kymlicka, Taylor, Modood, and a great number of those writing after are most critical of is what has been called 'benign neglect' or 'difference-blindness'; a position they ascribe to traditional liberalism. That is, the state, in both the setting up and enactment of institutions, laws, and policies should ignore the particular ways of life that its citizens follow, and instead focus only on some general characteristics of their citizenry; such as their shared fundamental interests. Their rejection of indifference, or what is usually understood as liberal neutrality, then allows difference-sensitive treatment; whether in the form of recognition, rights, or exemptions – in other words, respect for difference.

These are not the only charges against neutrality either. There are those who see 'neutrality' as simply masking and preserving the power of a dominant culture; that what is claimed as 'neutral' or 'indifferent' is in fact a status quo that seriously favours those with power and disadvantages minorities.

Yet the rejection of state neutrality and indifference has been far too quick and, in much of the multicultural and group-differentiated rights literature in particular, rests on several crucial misunderstandings. This means that the initial premise on which most of the argument for respect for difference rests fails: there is no need to contrast neutrality with difference-sensitivity, and certainly no need to assume that neutral and indifferent policies and institutions cannot fairly accommodate a substantial number and variety of ways of life. States will never be indifferent to *all* the ways of life within their domain – there remain cases where respect and forbearance tolerance is called for – indifference, in the form of neutrality, is the ideal, and, as I will argue here, is capable of a great deal of accommodation; and more so than respect for difference.

More specifically, against those who reject liberal neutrality, I make two claims: (1) neutrality is an ideal, it will never be fully realised, but still remains action-guiding; (2) neutrality is a range concept, and it thus needs to be sensitive to the changing nature of its particular range. These two claims in defence of liberal neutrality are themselves neutral between two rival conceptions of liberalism: one which advocates the state having little involvement in the cultural sphere and one which envisages a much more active liberal state. Neutrality may be aimed for by either actively assisting ways of life within a particular range (being 'hands on') or assisting none (being 'hands off'). This suggests that many of those arguing for a much more active state – and one which grants many of the claims of diversity – may be rejecting the name of neutrality but not its value.

Ultimately, though, if the aim of those who reject state neutrality in situations of diversity is to ensure the fair and equitable accommodation of a variety of divergent ways of life, then active or 'hands on' neutrality may not be ideal. This is my third claim: (3) neutrality is best realised by the state withdrawing support for favoured ways of life rather than by providing support for previously unrecognised or marginalised ways of life. This third claim acknowledges the epistemic and practical limits of modern states, and by not distinguishing between mere preferences and deeply held beliefs has the distinct advantage of avoiding the vexed issues of 'chosen-ness', 'authenticity', and so forth, which have plagued the debate over group-differentiated rights. Thus the main argument of the chapter is for the ideal of 'hands off' neutrality, understood as an active rather than passive form of indifference.

Neutrality as an ideal

The multicultural charges against neutrality and indifference have three distinct prongs. The first prong is that real liberal states *are not* neutral. That is, if we look at Canada for example, particular languages are privileged (English and French) and some – such as its Indigenous citizens – have particular rights that are not available to all citizens. This privileging of both majority and minority practices is not neutral or indifferent, but nor as the argument goes is it necessarily illiberal. Nor is Canada unique, most actual liberal states, albeit to varying degrees, do not simply 'stand back' in the cultural marketplace, but are very much involved in protecting and promoting various ways of life; be they languages, religions, or practices, and of both the majority and minorities.

The second prong of the attack is more theoretical; that is, liberal states *cannot* be neutral. For a state to function it will have to privilege some ways of life and not others. The most common example given here is language, with the argument that for a state to adequately function it needs a *lingua franca*. There are other areas that are commonly thought of as unavoidably non-neutral too. It has, for example, been argued that liberal democratic states require a particular political culture in order to function.[3] In even stronger form, liberal nationalists argue that states will require not just a generic liberal democratic culture, but a particular national culture whose accompanying feeling of togetherness provides support for public goods, especially redistributive justice.[4]

Accordingly, then, if liberal states *are not* and *cannot* be neutral, we should stop pretending they are or ever could be neutral. Further, if the good of neutrality lies in the freedom of citizens to follow their own ways of life, then, as the argument also goes, neutrality is failing and is better replaced by some non-neutral, or at least much less indifferent, form of state policy.

Yet we need not move so fast in the rejection of neutrality. The first problem with these two ways of rejecting neutrality is that they fail to see it as a political *ideal*. Ideals are ideals because they are never fully realised, both in practice and in ideal theory.

If we look at the ideal of democracy, for example, no contemporary state is democratic according to any of the various notions of ideal democratic theory out there. This does not mean that we cannot call particular states democratic, nor does it mean that we should abandon the concept of democracy altogether. What it does mean is that we can critique existing states using our particular normative ideal of democracy and encourage the state to come closer to this ideal while still balancing other values.[5] Neutrality here is no different: no actual state will be neutral, but this does not mean we cannot (nor should not) critique its practices against this ideal.

This balancing of other values leads to the more theoretical critique that liberal states *cannot* be neutral. Almost no ideal theory promotes a world of just one value. Even Rawls' promotion of justice as the ultimate value is one which already has other considerations (namely efficiency, security, and stability) built into it. Further, the sort of neutrality discussed here (that is, between ways of life) is not an ultimate value but a derivate value – other values are prior to neutrality and thus justify it, such as the freedom of individuals to live their lives as they see fit. As a derivative value it must be balanced by other considerations.[6] If, for example, the state really cannot provide a just distribution of resources without privileging a majority culture, then it should privilege that culture. This does not mean it can do anything to privilege this culture, but that it is justified in deviating from neutrality as much, and no more, than is necessary to achieve this outcome. Neutrality, like other ideals, particularly derivative ideals, needs to be balanced with other things that are held as important. As such, in both practice and theory it is not a reasonable critique of neutrality to say it cannot be realised.[7]

Yet ultimately, this is not the main criticism of liberal neutrality. While issues of feasibility are used to soften neutrality up, it is its desirability (or lack thereof) where the main blow is landed. This is the third prong: the liberal state *should not* be neutral in matters of difference as state neutrality either fails to lead to the accommodation of minority ways of life, or at least makes some minority ways of life comparatively and significantly more difficult to follow. It is this argument that liberal neutrality cannot adequately and fairly accommodate difference that I want to focus on in the rest of the chapter. I will argue that liberal neutrality has been wrongly characterised, and that a 'reflective', rather than a 'starting gate', notion of neutrality best encapsulates the ideal.

Conceptions of neutrality

Part of the maligning of neutrality seems to stem from some conceptual slipperiness, and people seemingly talking past each other.[8] So before putting the case for my own position, let me first briefly lay out the basic conceptions usually referred to in this domain. In this area of institutional neutrality, there are three basic conceptions that are discussed: neutrality of justification, of intent, and of outcome.[9]

Under *neutrality of justification*, no law or policy should be justified by the rightness of any particular way of life. The most famous example of this type of neutrality is Rawls' political liberalism, in which the principles of justice that regulate a polity must be consistent with public reason, that is, the form of reason that is not particular to any one conception of the good, and can reasonably be shared by all citizens.[10] Rawls' principles of justice are intended to have neutral justification; the rightness of no particular way of life is used to justify them, and they are intended to support the widest possible range of ways of life.

The second type of neutrality is of a different order. Here what matters is not so much how a law or principle has been justified, as much as its intention. Under *neutrality of intent*, a neutral institution or policy should not intend to favour (or hinder) any particular way of life. For example, a state that provided funding and support to opera but not pop music, or to Olympic rowing but not a rugby league, or to Catholicism but not Islam, would not be neutral in this sense. These types of policy decision intentionally help one way of life, and do not assist, and perhaps even hinder, its competitors. Neutrality of intent can be enacted through either act or omission. That is, a state can be neutral in this way by either standing back and doing nothing (what is sometimes called 'benign neglect'), or by offering all parties similar support, for example, by funding and supporting all religions in the same way. The former is a form of 'hands off' neutrality, and the latter a form of 'hands on' neutrality.

The third type of neutrality is effectively the flipside of neutrality of intent. *Neutrality of outcome* is concerned not only that institutions or policies do not intend to favour any particular way of life, but they do not actually favour – even unintentionally – any way of life. Thus, to use the opera/pop music example, a neutral (in intent) policy may support neither form of music. Yet such a policy is unlikely to have a neutral outcome. Pop music by its very nature is likely to be well supported, while opera is usually less popular and is comparatively much more expensive to produce. The *effect* on these two art forms would not be neutral. Thus, unlike the other two conceptions of neutrality, in the case of neutrality of outcome the state will most likely have to offer support to some ways of life; it is entirely unlikely that neutrality of outcome will be achieved by the state simply standing back.

It should be noted that *none* of these three conceptions of neutrality necessarily call for complete neutrality – in all three cases neutrality always has a distinct domain. A policy, for example, is neutral relative to a group of people and their ways of life. It is not, and could not meaningfully be neutral among all people and all ways of life. Even Rawlsian justificatory neutrality is not, in his own words, neutral in respect to everything. While it is neutral between conceptions of the good, it is not neutral about ultimate values (conceptions of the right) – and it is certainly not meant to be neutral to the 'unreasonable'. It is hard to see how any political authority could be totally neutral in this respect. It is important to stress this *range* aspect of neutrality. State institutions and policies will not be neutral in respect to *every* way of life but only neutral, at most, in respect to justice-respecting ways of life. Neutrality in all three conceptions is a range concept where everything within certain boundaries is to be treated neutrally.[11] Thus the common claim that 'there is no such thing as complete neutrality', while true, is only trivially so; no actual conceptions of neutrality strive for it, and it would be hard to envisage any remotely liberal political theory advocating such a position (even if it was somehow possible).

Of more interest is the likelihood of a policy or institution being neutral in one way but not in another. Thus a neutral justification may lead to a policy that has non-neutral intent. And a policy of neutral intent may lead to non-neutral outcomes. For example, an official state language could be justified neutrally. The existence of a *lingua franca* might be justified on grounds of efficiency, social cohesion, and perhaps as a prerequisite for social mobility.[12] According to these potential justifications, polities function better, people feel more connected, justice can be both served and seen to be done, and people may be more likely to avoid being 'victims' of their birth when there is a core language.[13] While these potential justifications are neutral, they will certainly not lead to neutrality of intent – indeed quite the opposite as one language is to be promoted ahead of others. Whichever language is chosen, it will unlikely be as easy for every citizen to converse in it. Indeed, as already mentioned, this is one of the charges against the non-neutrality of actual liberal states. As far as neutrality of outcome is concerned, we do see in practice that when particular languages are promoted, other minority languages tend to die off.

It should also be clear that neutrality of intent will commonly be in tension with neutrality of outcome; giving equal resources to all religions, for example, does not mean they will be equally successful. More interestingly though, is the way that over time neutrality of intent may fail on its own terms. If we take the example of religious holidays, we can imagine a society where several different denominations of Christians reside, perhaps even with quite a high degree of animosity, and all of which recognise Sunday as the day of rest. The state enacts a policy that formally recognises this as a day of rest and does so with neutral intent – it does not intend to favour any of these competing ways of life (and indeed the outcome is neutral too). Yet over time, other religions with different days of rest begin to be practised in this society and some Christians abandon religion altogether. This policy, while being neutral in intent when first implemented, is now very biased towards particular ways of life and there is a good case that it unfairly impinges on some individual's lives more than others.

For the liberal concerned with dealing fairly with diversity, there are now two possible problems with neutrality. First, neutrality of justification can come into tension with neutrality of intent, and, second, neutrality of intent may, over time, take us a long way from what we may commonly consider neutral treatment. This would mean that as its critics charge, though not necessarily for the same reasons, these two traditional understandings of neutrality may not be adequately and fairly accommodating of diversity.

Active indifference

My aim here is to show how neutrality can be made sufficiently and fairly accommodating, and in the process rebuff the argument that liberal neutrality cannot adequately deal with diversity. My point here is to argue on the terms of (at least most of) the critics of neutrality; and to show how it is neutrality that brings about fair and equal accommodation of difference, and further that a commitment to fair and equal accommodation is actually a commitment to a principle of neutrality.

What would it mean, then, to be guided by the ideal of neutrality within a particular range? In brief, it would mean consciously trying not to favour any particular ways of life. Of course one way to do this would be to somehow look into everybody's life plans and desires and then try and fulfil them. This would be a complete neutrality of outcome. The most basic problem with this approach is that it would be apolitical. If it really was possible for all people to do what they wished (even if we exclude things that harm others), then politics would have no role. Neutrality of outcome may be relevant in a limited domain (perhaps between a limited number of discrete things, such as organised religions), but it is not relevant on a general scale. There are also good reasons for not using neutrality of outcome as an unrealisable but still action-guiding ideal: it fails a basic choice-sensitivity test and may undermine any individual responsibility for life-choices.[14] This leaves us with neutrality of justification and neutrality of intent.

I will accept the importance of neutral justifications here (that is, not using the rightness of any particular way of life to justify a law, policy, or institution). But, importantly, this does not necessarily rule out promoting particular ways of life. As mentioned, a value, such as justice, *may* actually require the promotion of a particular way of life (this is the liberal-nationalist argument). Or another set of values, such as those of efficiency, social cohesion, and social mobility, might dictate the need for the promotion of a national language. In these cases neutral justifications have been given for policies which have non-neutral intent. But this does not mean neutrality of intent no longer plays a role. If we understand it as an action-guiding ideal, rather than a binary property, then neutrality of intent is still relevant. So if, for example, a common language is required, then its choice and use needs to be implemented to just the level required to achieve its goal and no more. It is hard to say in the abstract how far this should be but the modern European history of nation-building, which commonly involved the deliberate wiping out of regional dialects and the punishment of those who continued to speak them would certainly be going too far.[15] The ideal of neutrality between ways of life needs to be balanced with the particular policy goal (even if justified neutrally) to thereby keep it in check. If, for example, there are sufficient reasons of social coordination to favour one way of doing things over others, then on a neutral justification this is permitted. Neutrality of intent is not then nullified, but remains a balancing value.

Let us turn directly, then, to neutrality of intent, which seems the most relevant in this context. The major problem with this type of neutrality was that it seemed very static: a policy, institution, or law is created, and what matters is that it does not intend in its creation to favour any particular way of life. But if neutrality is to be an action-guiding ideal, then this is not sufficient. Because neutrality is relative to a range, it needs to be responsive to any changes in this particular range. So a policy on religious practice needs to be sensitive to the emergence of new religions, and not just those that were present when the policy was first designed. This means that if there are new ways of life within its domain, maintaining neutrality may well require changing a particular policy, institution, or law.

This balancing act needs to be ongoing rather than simply one-off. A requirement for police officers to wear a uniform could have been both justified neutrally, and initially implemented with neutral intent in relation to the ways of life of a particular state's citizens. That is, upon implementation, the uniform is one which is not problematic for any particular way of life. However, if the freedom of individuals to live their lives as they see fit is taken seriously, this cannot be the end of the story. People change, fashions and tastes evolve, and migrants with different ways of life enter a political community. The existing uniform regulations may become much less neutral than was initially intended – those who usually wear religious garments, for example, may feel they cannot join the police force. A form of neutrality that treats fairly the freedom of individuals to live their lives as they see fit needs to be able to respond to changing circumstances and change its policies to maintain a more commonsense neutrality. This is *not* neutrality of outcome, but nor is it neutrality of intent as usually understood. It is instead a realisation that a policy that is both *knowingly* and *avoidably* not neutral in intention *at that point in time* should to be changed. This is not a fourth category of neutrality,[16] but is neutrality of intent as it *should* be understood – that is, as active and responsive, rather than simply passive.

This notion of state neutrality is range-sensitive. An agent is never neutral in the abstract but instead neutral among certain things; in this case, ways of life. On this understanding of neutrality, the state needs to be sensitive to the changing nature of the things it is neutral among. In the police uniform example, this neutrality can be realised by expanding and changing the range of possible uniform variations: skirts as well as pants, long sleeves as well as short ones, alternative headwear in the same colour as the existing hats etc. This conception of neutrality takes seriously the idea of 'intentionally not favouring' any particular ways of life by applying this condition beyond simply the design and setting up of policies and institutions.

Here, then, is a conception of neutrality that takes seriously the fact that neutrality is a range concept, and is in fact, 'difference-sensitive'. It acknowledges that the range of differences over which the state is neutral may expand and adapt – the details of the range of tolerable differences cannot be decided once and for all at a single point in time, and the details of what occurs within that range cannot be known ahead of time. While in justificatory neutrality, at least of the Rawlsian variety, this range manifests itself through the *exclusion* of ways of life altogether such that only general principles are permitted, in the case of neutrality of intent, the intention is not to privilege any *actual* ways of life. While this may be moderately easy to do when a policy or institution is first implemented – there will be an existing range of ways of life – over time this range will change and so too must the policy if it is to remain neutral in this way.

In the police uniform example, the policy of uniform wearing is feasibly justified on the grounds that police need to be recognisable, and that a common uniform helps provide some discipline for them as well as authority in the eyes of the general public. These are seemingly neutral justifications – and recall I am assuming neutral justifications are necessary. Yet these goals can be achieved with some variation (as now happens in most Western liberal multicultural jurisdictions), and thus there is insufficient justification for the uniform to be strictly uniform, so to speak.

I should stress that in some cases the initial goal itself may need revision, particularly if it is not reasonably justified neutrally. That is, the exposure as being non-neutral in intent may highlight a deeper and unjustified non-neutrality, and one which certainly challenges the status quo. Policies and practices need neutral justifications, but properly neutral justifications can allow a wide range of practical possibilities.

Why 'hands off' is better than 'hands on'

The argument that liberal state neutrality should be thought of as dynamic and difference-sensitive is consistent with two forms of liberalism: that of recognition and an active state, and that of withdrawal and a smaller state. What I will call 'difference-sensitive neutrality' or 'active indifference' can be realised by the state either withdrawing support for privileged ways of life (being 'hands off'), or maintaining this support and extending the privilege to the previously excluded (being 'hands on'). In this section I want to add a third element to this conception of neutrality (in addition to it being an ideal and responsive to range), which brings it much closer to a more traditional 'hands off' understanding of neutrality. This involves a strong preference for moving closer towards neutrality by removing rather than adding support in situations where non-neutrality needs addressing.

Anna Elisabetta Galeotti, in her 'toleration as recognition', puts forward an example of a 'hands on' version of neutrality which has similar features to the understanding of neutrality I have been advancing.[17] Galeotti argues that when it is realised that the state is acting non-neutrally, the state should not simply expand its range of neutrality, but that, in so doing, it should also publicly declare the previously excluded way of life to be of equal worth. That is, because some ways of life were previously privileged, the excluded ways of life should now also receive privileged treatment. This approach, like other forms of active recognition or respect for difference, does not advocate removing recognition or privileged status but adding new ways of life to the list. In this sense neutrality is still the intent, and so this is a conception of neutrality that is also sensitive to difference.

Yet, as I want to show, 'hands on' forms of difference-sensitive neutrality are generally inferior to a more 'hands off' form of neutrality and its strong preference for withdrawing support rather than offering support when it is realised that neutrality is being unjustifiably violated. A couple of examples may help draw this out. In New Zealand, elections are held on Saturdays. Saturday polling may seem to unfairly and avoidably impact on some citizens, including religiously orthodox Jews, whose Saturday Sabbath takes the day of rest very seriously. One option would be to exempt religious Jews from Saturday polling, either entirely, or to give them another option, Sunday perhaps. This would be to actively recognise these Jews and offer a neutralising alternative. Yet neutrality can be achieved in another way which does not recognise any particular group or grant exemptions, and also allows other individuals greater freedom. In New Zealand, the alternatives of pre-poll and postal voting demonstrate both that standing back is possible, and allow any citizen to choose to vote prior to polling day. These are equally valid options for someone who will be away fishing, travelling, or working a double shift, as it is for someone with a deeply held religious belief, and they provide greater freedom for all these people. Another option, of course, would be to try and choose a different day altogether, say a Tuesday, on which no one had their Sabbath. But this last option, while perhaps neutral in intent, would seemingly fair worse on a neutral justification: maximising the number of people who vote.[18]

'Gay marriage' debates have been going on with various degrees of success across almost all Western states and provide another interesting case. One of the main arguments given by advocates of gay marriage is that because the state recognises heterosexual marriage, it is unfair to homosexuals that their way of life is not equally recognised (both symbolically and/or legally). In other words, it is a non-neutral policy. Those who argue for recognition or for 'hands on' neutrality are seemingly committed to offering homosexual couples the option of marriage.

But there are two other alternatives. The first would be simply removing the reference to 'between a man and a woman' from the relevant marriage act. This is what many gay marriage advocates are actually arguing for, and this move simply makes sex and gender irrelevant to marriage. The second would involve withdrawing the recognition of marriage by the state altogether; with marriage becoming an entirely private religious or cultural matter.

With all three alternatives, the state would recognise that it is avoidably, and seemingly unjustifiably, privileging one way of life over another and thus needs to become more neutral. Recognising gay marriage, opening it up altogether, or entirely withdrawing marriage as a state issue, can be seen as moving closer to the ideal of neutrality. So which is the more favourable approach?

To answer this, we can return to the shared premises underlying this dispute: the importance of living ones' life as one sees fit and treating these ways of life fairly. A difference-sensitive neutrality that removes favour rather than adding new classes of favour is more likely to increase the space for individuals to realise this freedom. While recognising gay marriage is better than only recognising heterosexual marriage, it is still not as neutral as could be. If the aim is to be neutral towards people's significant relationships, then expanding the options from one to two still excludes and thus treats potentially unfairly other forms of legitimate and important relationships. The form these relationships could take will be difficult to list and then recognise in the abstract – surely there are many possible variations. It is much better for the state to simply treat whatever form of relationship people have in the same way, rather than privileging only some recognised forms. This is also the problem with simply amending the relevant marriage act by removing the reference to 'between a man and a woman'. In Western societies, marriage, both as an institution, and in the language the state uses to describe and regulate it, privileges particular kinds of relationships – those that involve sexually monogamous romantic love. Privatising marriage, on the other hand, would not pre-judge which form of relationship individuals hold as the most significant for them. It would also allow particular religious groups to put quite restrictive parameters on marriage for their members but for those members to easily leave their group if they so wish.

One can go further here. In privatising marriage, the state is withdrawing from the more symbolic and normative aspects of marriage, but the practical elements of people's relationships will still likely need state regulation. It matters, for example, who has rights in a child custody dispute, or who gets what when someone dies without a will, or how people are treated for taxation and welfare purposes. So what should a neutral state do? One option is to provide an alternative form of contract which does not privilege romantic and exclusive love. The *Pacte Civil de Solidarité* (PACS) in France provides a rough example of this in practice. While initially designed for same-sex couples and with a lesser set of rights, it has mainly been taken up by heterosexual couples, and, legally speaking, is now very close to marriage (just missing full inheritance and adoption rights).[19] There is nothing in the PACS about romantic love or sexual exclusivity, and it is open to any two people who are not bound by another PACS a marriage, or who share lineage.

To use Clare Chambers' term, the PACS is an example of a holistic approach to regulating relationships; people must opt-in and then they receive a bundle of rights.[20] But, as Chambers argues, while this is preferable to state-sanctioned marriage, it is not without its problems.[21] The fact that people must opt-in can leave vulnerable those who are living in functionally (but not legally) equivalent relationships.[22] A good example is an uncontracted couple with children who have taken on traditional gender roles. In the event of a split, the primary carer may be left significantly worse off. The holistic approach also assumes people will get all the relevant functions within one relationship. It does not, for example, capture separated couples who co-parent, or people who may have both a traditional nuclear family and share property or care responsibilities with extended family members. This second problem is simply a restatement of the problem that the state cannot easily capture people's significant relationships, and should not, at least if it is to be neutral, assume a norm. It may be that complete neutrality is not possible here – and since neutrality is an ideal, this is certainly not fatal – but this discussion does suggest the importance of not trying to prescribe which type relationships will be the most significant, and having as wide a range of neutrality as possible.

One of the main reasons for favouring 'hands off' neutrality is that it offers more freedom to all of us. It does not require individuals to be recognised as belonging to a particular group – with all the attendant problems of 'authenticity' – nor does it require political advocacy, which is the usual requirement of most new forms of recognition. It also allows individuals to freely experiment with ways of life. If, for example, a 'burqini' is permissible as part of a standard beach or pool life-guard uniform, then anyone can try it out – they need not first become a recognised 'authentic' Muslim. We can imagine such a uniform option might be attractive to a number of people, not just those who are either considering or practising orthodox Islam.

In all these cases, the existence of gay marriage advocates, or religious Jews wishing to exercise their civic duty, or orthodox Muslims wishing to act as life-guards act as triggers for re-evaluating the neutrality of a particular policy or institution. Yet in its response, the state should not just respond singularly to claims of non-neutrality. In its review of existing policies it should try to find the most neutral solutions, and ones which are potentially neutral between more than just the current privileged way of life and those who are making a claim of non-neutrality. While those who make claims of non-neutrality upon the state often have genuine and important demands, it is likely that there are those less eloquent who also have important claims, and perhaps even those who are yet to realise or worry about whether their way of life is treated non-neutrally. Indeed the gay marriage issue seems to be partially one of 'not yet realising'.[23] The occasions when the particular policies and institutions of modern states change are quite rare. By removing privilege on these occasions, the new policies and institutions are likely to be neutral among a greater range of ways of life than if new kinds of recognition are simply added. In other words, the acknowledgement of avoidable non-neutrality in a particular area provides a unique opportunity for a modern state, and one that should be taken full advantage of. Difference-sensitive neutrality or active indifference thus takes seriously that not only is the modern state a lumbering beast which changes very slowly, but also that it is unlikely to have good knowledge of people's actual ways of life, nor predict how they will evolve in the future. This is in contrast to those who call for more activity and recognition, who seem to assume a state which has excellent knowledge, is dynamic, can change very quickly, and more importantly, do all this correctly. Thus the primary reason for 'hands off' rather than 'hands on' neutrality is based on a view of what the state can and should do given its limitations.

At this point, it might be objected that active indifference is really just respect for difference in disguise. That is, in making changes to accommodate particular practices, the real motivation is to somehow respect these practices. As I noted, it is true that the politics of such changes will likely be motivated by vocal claims for accommodation from or on behalf of particular sections of society. But an acknowledgement of the politics of change is different both from saying this is the sole motivation for change, nor that such change actually respects particular differences. Yes, active indifference in seeking to expand the range of neutrality acknowledges the presence of such claims, but in removing unjustified privilege there need be no intention to favour these claims, which can simply be the triggers for revisiting the neutrality of a particular policy, law, or institution.

Not distinguishing between mere preferences and deeply held beliefs

Given the size of modern states, and the almost infinite amount of often changing diversity, it seems that trying to be 'hands on' will be more difficult than being 'hands off', especially if one is concerned with protecting individual freedom and not damaging identity. It seems that most of the time it will be easier and more viable to remain neutral by withdrawing support rather than constantly trying to juggle support for various minorities.[24]

It may appear that this fluid and dynamic picture of the various ways of life within a polity also poses problems for my conception of difference-sensitive neutrality or active indifference; it too needs to respond to the diversity that actually exists. Yet by expanding the space for diversity, difference-sensitive neutrality does not just respond to the initial potential trigger of non-neutrality. While these triggers, such as the gay marriage advocates, inform the state of their non-neutrality, the state's response of withdrawing support for particular ways of life makes the policy or institution more neutral between actual ways of life. Yes, it might not always get it right, but by not recognising any particular way of life it is much less likely to fix or distort an identity, trap people within their groups, or inhibit people's legitimate ways of life. Of course, not all ways of life will succeed, but if this is the yardstick, then as I mentioned earlier, politics would have to be transcended.

To what extent, then, is this understanding of neutrality multicultural? It certainly does not prescribe identity recognition, exemptions from general laws, or group-differentiated rights. But nor does it prescribe assimilation or a privileging of majority ways of life – indeed quite the opposite. Difference-sensitive neutrality is not anti-multicultural, but at the same time does not support claims for diversity, and instead simply tries to make space for them.

Where this conception of neutrality also differs from the usual conceptions of multiculturalism is that it does not try to distinguish between types of difference. Its intention is to be neutral between mere preferences *and* deeply held beliefs, and not try to dig too deep or use some criteria as to why some people's differences may matter more than others. In this sense, there is nothing special about religious diversity. Expanding the range of neutrality, but not granting exemptions or recognition allows this lack of distinction – and this is a significant and important advantage. First, by avoiding the need to distinguish between types of difference, it side steps the vexed questions of 'authenticity', 'chosen-ness', and 'alterability' which come with positive recognition. It also avoids the epistemic problem for decision-makers of both understanding someone's difference and its importance. Likewise, it may help minimise the construction or at least emphasis on particular kinds of difference – such as 'sincere', 'conscientious', or 'religious' belief' – which are more likely to hold political sway. Second, for those, such as autonomy-liberals, who think it matters that we have more rather than less viable options, it more easily allows 'experiments in living'. Recognised exemptions and group-differentiated rights require officials to decide who is eligible and who is not, and thereby limit the flexibility of individuals to choose for themselves which identities or practices to subscribe to. With the conception of active indifference that I have put forward, it is entirely up to the individual which way of life he or she identifies with, and state officials do not need to examine their reasons. So, for example, if a uniform changes to allow a variety of headwear, then it is up to individuals which type of headwear they quite literally 'try on', and not a matter of being eligible (or not) for a recognised exemption.

Where numbers and intensity will likely matter, however, is when it comes to triggers for expanding the range of neutrality. Because neutrality is but one value that needs to be balanced with others, the way the case for change is made is likely to matter politically. A significant number of people who claim to be treated non-neutrally and in a significant fashion, is likely to affect whether or not the range of neutrality is changed. This is not to say these things should matter in a pure conception of difference-sensitive neutrality with some omniscient state, but to realise that changing policies and institutions is difficult and should not be done lightly. Yet, whatever the trigger to expand the range of neutrality, the process of change should look beyond it to other potential ways of life that may also be being treated non-neutrally.

Let me clarify this point a little further. My claim is that there is no reason in themselves to favour making space for conscientious beliefs over preferences when a law, institution, or policy unjustifiably does not permit certain actions. Both provide pro tanto justifications for change. If both can be accommodated and a neutral justification still met, then they both should be – religious or conscientious beliefs should not be the only reason to change a policy. Returning to the police uniform example, if a neutral justification can still be met, the claims of the fashion-conscious and the claims of turban-wearing Sikhs should be accommodated. Where a line may have to be drawn is where only one claim rather than both can be met. In these cases, other considerations may have to come into play. So if, for example, the Sikh community have a long history of marginalisation or perhaps tension with the police, that may favour turbans. But if, on the other hand, police numbers are low and making the uniform more fashionable would greatly increase recruitment, then this may favour the fashion-conscious.[25] And here the strength of the claim may also have to be taken into consideration – although it should not be assumed this will always favour religious-type practices. Finally, if it is not already clear, by preferences I am not including ones that are primarily other-regarding. While the distinction between self- and other-regarding preferences is notorious, I think there are meaningful cases on both sides. Wanting not to eat meat is the sort of preference that should be taken seriously, but wishing other people would not wear short skirts is not.

While the theoretical ideal of difference-sensitive neutrality involves a pulling back of unjustified non-neutral support, this ideal may not always be realised in practice. This can be for several reasons. Sometimes it may be impractical – it might, for example, really be better to have only a small number of national languages. At other times there may be negative consequences of changing the status quo – Brian Barry suggests this in relation to removing what he sees as unjustified exemptions from humane animal slaughtering laws for Jews and Muslims.[26] At other times democracy may be a barrier, and it might not be politically feasible to reverse some areas of favouritism – subsidies for farmers or for Olympic sports are possible examples here.

In these cases, the lack of the real possibility of removing support for particular ways of life does not mean that nothing should be done. In the first place, as I argued earlier, support should be reduced as much as is feasible in order to come closer to the ideal of neutrality. Failing this, there may be occasions where taking a more 'hands on' approach is the right thing to do. For example, in the marriage case, if it is not democratically possible to remove state support for heterosexual marriage, then homosexual marriage should be recognised by the state. This position would be less neutral than removing support altogether but more neutral than only supporting heterosexual marriage, and is thus to be preferred.

Finally, there is the issue where the non-neutrality of institutions has been so egregious that withdrawing support for the previously favoured is not enough, and some sort of redress is justified. Support for minority languages and affirmative action policies along race or gender lines are sometimes justified in this redressive fashion. If justified, then this is no longer neutrality of intent but a form of neutrality of outcome. Because such policies are redressive, they should only ever be transitional – once the wrong is redressed, they are no longer justified. This means that active indifference as 'hands off' remains the ideal – not only should the policy be temporary, but its implementation should be constrained by this ideal. That is, if there are ways of achieving redress which minimise recognition, then these should be preferred, especially as it is very difficult for such policies to avoid the problems of state recognition of difference.

Let me stress that my aim here is to show the possibilities and strength of a particular approach to the issues on contemporary diversity. I do not claim that no exemptions and the like can ever be justified, just that there are good reasons not to grant exemptions and sound alternatives to doing so. To use a common example, my aim is not to say definitively whether or not Sikhs should be allowed to carry knives, but that there are strong arguments for either everyone or no one being allowed to carry knives. The debate should be about neutral justifications; as I have argued, once these are settled, one group should neither be exempt nor favoured by their application. But as I have also argued, there is normally more than one way to realise a neutral justification, and the way that is most neutral to the ways of life that exist should be favoured. Nevertheless, this is a pro tanto judgement, and real-world politics among other considerations may ultimately force the acceptance of a second-best option.

Conclusion

At the start of the chapter, I noted that many of those who argue for state respect for difference used the failure of neutrality as a crucial first premise. Their claim was that because liberal states were not and/or could not be neutral, and, more powerfully, should not be neutral, a variety of differentiated treatment, or respect for difference, could be justified. In response, I have argued that neutrality is an unrealisable, yet action-guiding ideal that necessarily needs to be balanced with other values; that neutrality is a range concept, and that neutrality of intent, in particular, needs to be sensitive to the changing nature of this range; and, finally, that the values that justify neutrality are best realised by a 'hands off' rather than 'hands on' form of neutrality. Accepting this argument pulls most, if not all, of the rug out from under multicultural and group-differentiated rights theory, and with it the claim that states should respect difference: the liberal state can have a policy of neutrality or indifference and still fairly and reasonably accommodate a variety of ways of life, and respect of difference is not necessary. In achieving this accommodation, the vexed questions of who, when, and how various identities/practices/beliefs should be recognised are almost entirely avoided. When the state is the agent of toleration, indifference, and not respect for difference, should be the preferred pathway to accommodation.

Notes

1 Balint (2017); King (1976, pp. 12–13, 68). See also Horton (1996, p. 38). This chapter draws heavily on Balint (2017), Ch 3.
2 In particular, Kymlicka (1995); Taylor (1994); Modood (2007).
3 Bader (1997).
4 See, for example, Miller (1995); Tamir (1993); Soutphommasane (2012); and Lenard (2012).
5 While 'democracy' can mean 'all good things', this is not the sense I mean here. Most theoretical conceptions of democracy are more precise and do not encapsulate all ideals.

6 Patten (2014, pp. 108–111), and Patten (2012), similarly characterises neutrality as a 'downstream' value which provides a pro tanto constraint on action, while Goodin and Reeve (1989) describe it as an instrumental rather than fundamental value.

7 Cf. Carens (2000, pp. 11, 52–87).

8 Cf Gaus (2003, p. 138).

9 Different authors use different labels for these three conceptions, and some also distinguish different conceptions of neutrality. See, for example, Marneffe (1990); Kymlicka (1989); and Patten (2012). I have chosen these three as they are among the most common (Arneson, 2003, p. 193) and allow me to focus the discussion of what seem to be key issues for neutrality and diversity.

10 Rawls (2005).

11 By 'range concept' I do not simply mean 'context-dependent'. Although context will often help set the appropriate range, it is also possible that the range of neutrality could be set by first principles.

12 Barry (2001).

13 South Africa, with 12 official languages, provides an obvious exception to this line of reasoning.

14 Kymlicka (1989, pp. 884–885). Cf. Patten (2014, pp. 137–147).

15 For the French case, see Weber (1977).

16 This is in contrast to Patten's (2012, 2014) 'neutral treatment'.

17 Galeotti (2002).

18 This is of course assuming that people are less likely to vote on work days than non-work days.

19 By 2013, 96% of all PACS unions were heterosexual. Mazuy, Barbieri, and d'Albis (2014, p. 288).

20 Chambers (2013).

21 It is worth noting that having standardised contracts of this sort is likely to be better than complete privatisation which may leave poor people vulnerable as they may not be able to afford better quality (and more protective) contracts. I thank Kerri Woods for this point.

22 Chambers (2013, pp. 135–137).

23 See Altman (2011) for the historical antipathy towards marriage by homosexual advocates.

24 Patten's (2014, 2012) recent argument for 'neutrality of treatment', and with it the justification for 'minority cultural rights', has interesting similarities (but important differences) to the argument I have pursued here. Patten characterises 'neutral treatment' as being 'equally accommodating' of different conceptions of the good. This can be achieved by either 'privatisation' (akin to 'hands off'), 'evenhandedness' (akin to 'hands on'), or 'generic entanglement' (where the state provides a good that can be shared by various conceptions of the good – such as a multi-sports facility). Depending on what the range is specified at, it seems this third option can in some cases be 'hands on' and in others, 'hands off'. While Patten acknowledges the problems with 'hands on' neutrality (2014, pp. 122–123), he ultimately uses it to argue for minority rights. Unlike Patten, I do not see 'evenhandedness' as ideal (for the reasons mentioned in the text), and certainly not capable of grounding rights. But methodologically, we both view neutrality as a pro tanto value, and thus it may well be the case that many of our policy prescriptions will be the same – I do not argue that 'hands on' neutrality should not be pursued, just that it is not as good as an active and reflective 'hands off' form of neutrality. Finally, as mentioned, Patten sees 'neutral treatment' as a distinct fourth category of neutrality. My 'difference-sensitive neutrality' is not a new category, but simply a full and proper understanding of neutrality of intent.

25 The use of fashion designers to design Italian police uniforms would seem a potential example here.
26 Barry (2001).

References

Altman, D. (2011). Same-sex Marriage Just a Sop to Convention. *Australian Literary Review*, 2 Feb.

Arneson, R.J. (2003). Liberal neutrality on the good: An autopsy. In: G. Klosko and S. Wall, eds., *Perfectionism and neutrality: Essays in liberal theory*. Lanham, MD: Rowman& Littlefield, pp. 191–218.

Bader, V. (1997). The cultural conditions of transnational citizenship: On the interpenetration of political and ethnic cultures. *Political Theory*, 25(6), pp. 771–813.

Balint, P. (2017). *Respecting toleration: Traditional liberalism and contemporary diversity*. Oxford: Oxford University Press.

Barry, B. (2001). *Culture and equality: An egalitarian critique of multiculturalism*. Cambridge: Polity Press.

Carens, J. (2000). *Culture, citizenship and community: A contextual exploration of justice as evenhandedness*. New York: Oxford University Press.

Chambers, C. (2013). The marriage-free state. *Proceedings of the Aristotelian Society*, 153(2), pp. 123–143.

Galeotti, A.E. (2002). *Toleration as recognition*. Cambridge: Cambridge University Press.

Gaus, G.F. (2003). Liberal neutrality: A compelling and radical principle. In: G. Klosko and S. Wall, eds., *Perfectionism and neutrality: Essays in liberal theory*. Lanham, MD: Rowman & Littlefield, pp. 137–166.

Goodin, R.E. and Reeve, A. (1989). Liberalism and neutrality. In: R.E. Goodin and A. Reeve, eds., *Liberal neutrality*. London: Routledge, pp. 1–8.

Horton, J. (1996). Toleration as a virtue. In: D. Heyd, ed., *Toleration: An elusive virtue*. Princeton, NJ: Princeton University Press, pp. 28–43.

King, P. (1976). *Toleration*. London: George Allen & Unwin.

Kymlicka, W. (1989). Liberal individualism and liberal neutrality. *Ethics*, 99(4), pp. 883–905.

Kymlicka, W. (1995). *Multicultural citizenship: A liberal theory of minority rights*. Oxford: Oxford University Press.

Lenard, P.T. (2012). *Trust, democracy, and the multicultural challenges*. University Park, PA: Pennsylvania State University Press.

Marneffe de, P. (1990). Liberalism, liberty, and neutrality. *Philosophy & Public Affairs*, 19(3), pp. 253–274.

Mazuy, M., Barbieri, M. and d'Albis, H. (2014). Recent demographic trends in France: The number of marriages continues to decrease. *Population* (English edition), 69(3), pp. 313–363.

Miller, D. (1995). *On nationality*. Oxford: Oxford University Press.

Modood, T. (2007). *Multiculturalism*. Cambridge: Polity Press.

Patten, A. (2012). Liberal neutrality: A reinterpretation and defense. *Journal of Political Philosophy*, 20(3), pp. 249–272.

Patten, A. (2014). *Equal recognition: The moral foundations of minority rights*. Princeton, NJ: Princeton University Press.

Rawls, J. (2005). *Political liberalism: Expanded edition*. New York: Columbia University Press.

Soutphommasane, T. (2012). *The virtuous citizen: Patriotism in a multicultural society*. Cambridge: Cambridge University Press.

Tamir, Y. (1993). *Liberal nationalism*. Princeton, NJ: Princeton University Press.

Taylor, C. (1994). The politics of recognition. In: A. Gutman, ed., *Multiculturalism*. Princeton, NJ: Princeton University Press, pp. 25–74.

Weber, E. (1977). *Peasants into Frenchmen*. London: Chatto & Windus.

3 Toleration and tolerance

Between belief and identity

Peter Jones

How much space should there be for toleration in a plural society ideally conceived? The unsurprising answer would be, 'a great deal'. Difference and plurality make toleration both possible and desirable. Hence, in circumstances of plurality, the tolerant society is generally applauded, as is the tolerant person. There are, of course, things that we ought not to tolerate and therefore things of which we are rightly intolerant, but, in spite of that truth, 'tolerant' is still commonly a term of commendation and 'intolerant' a form of reproach. As the pluralism that characterises contemporary societies has increased, so politicians and social commentators have become ever more vocal in urging the merits of toleration. Yet not everyone joins in that chorus of approval. Toleration may frequently be preferable to intolerance, but toleration is for some commentators a tarnished ideal and perhaps no ideal at all; it falls significantly short of the socio-political condition to which the members of a plural society should aspire.

Two features in particular, which are fundamental to the traditional idea of toleration, give rise to these misgivings. One concerns power. I tolerate your conduct only if I have the power to prevent or impede it if I choose. If I lack that power, I am in no position either to tolerate or not tolerate your conduct. Thus, I tolerate your conduct only if I possess the power *not* to tolerate it. Toleration may therefore seem to entail an asymmetry of power. I can tolerate you because I have power over you, but you cannot tolerate me because you have none over me. That picture of superordinate and subordinate maps easily onto the political world of the post-Reformation era, in which rulers, if they tolerated those who dissented from the officially approved faith, did so as a matter of grace and favour. It is at odds with the equal status of citizens that we associate with a democratic society. It also inverts the idea of indirect democracy, according to which an elected government stands to its electors as servant rather than master. Perhaps, then, toleration is an ideal that belongs to the past and little if any space remains for it in a democratic age (cf. Heyd, 2008; Newey, 1999).

That is not an objection that I shall try to counter here. I observe only that when people speak of a 'tolerant society' nowadays, they typically have in mind a horizontal model, in which the members of a plural society extend toleration to one another, rather than a vertical model, in which some as a matter of grace and favour hand down toleration to others. Of course, democratic arrangements do not guarantee toleration of that horizontal sort. A majority may have the whip-hand over minorities and, if it extends toleration to them at all, it may do so in the same condescending spirit as an early modern monarch. But toleration in a democratic society does not have to take that form. It can be part of a society's shared commitment to the right of its citizens to live the different forms of life to which they are committed, a freedom which each should enjoy on terms equal with everyone else.

A second feature that frequently prompts doubts about toleration is its negativity. We tolerate only that to which we object. If we do not object, we have no occasion to tolerate. Dislike or disapproval is therefore an essential feature of toleration. Yet nowadays we are commonly enjoined to cherish and celebrate the pluralism that characterises contemporary societies. We should view diversity or difference positively (cf. Addis, 1996; Beneke, 2006; Brown, 2006; Brown and Forst, 2014). If we regard some differences with dislike or disapproval, we should seek to overcome our negativity, which is to say that we should aim to move 'beyond' toleration to a positive outlook in which we accept, extend recognition to, or positively endorse that to which we currently take exception. Thus, even if a society's toleration does take a horizontal form, it may still fall short of the ideal for which we should strive.

It is that objection to toleration that provides my point of departure in this chapter. The chapter unfolds as follows. In the first section, I consider toleration in the context of different and conflicting beliefs and suggest that it is idle to object to the negativity of toleration in that context. In circumstances of disagreement, mutual toleration is not second-best but the best there can be. Nowadays, however, the differences that are salient for plural societies are commonly construed as differences of identity. In the second section, I detail several reasons why the toleration that seems so desirable for differences of belief can seem decidedly inappropriate for differences of identity. Properly conceived, identities should not require toleration. Before looking at some of the complications that arise from toleration's relating differently to beliefs and identities, I pause in the third section to consider whether tolerating must entail negativity. Several commentators suggest that it need not. In particular, 'being tolerant' is often used nowadays to describe refraining from disapproval or dislike, rather than refraining merely from the suppression of what we disapprove of or dislike. I use the term 'tolerance' to badge that notion of being tolerant. Tolerance, so understood, may be an entirely acceptable response to differences of identity.

In the fourth and fifth sections, I comment on the way in which the difference between beliefs and identities bears on the case for toleration in two areas of controversy concerning religion. In the fourth section, I consider 'challenges' to a religion – by which I mean episodes such as the Rushdie Affair, the Danish cartoons of Muhammad, and the representation of Jesus Christ in *Jerry Springer the Opera* – and how our conception of what is at stake in those challenges depends on whether we accentuate belief or identity. In the fifth section, I consider the accommodation of religion – a society's making special provision for the demands of religious faith – and consider how the belief/identity distinction bears on the case for that accommodation.

The toleration with which I shall be concerned is public rather than private in that it is toleration amongst the citizens of a society rather than toleration shown by individuals in their personal relationships with one another. Nevertheless, that public toleration is horizontal rather than vertical in form in that it is toleration amongst those who make up the citizenry of a society. Some of what I say relates to the rules that should obtain in a plural society but those rules remain horizontal in orientation in that they govern how citizens are to relate to one another. They do not manifest either toleration or intolerance by rule-makers of those to whom their rules apply.

Toleration and belief

In European societies during the modern era, the paradigm case of toleration has always been religious toleration. Why? We might suppose that the answer lies in the way that religion divided post-Reformation European societies both within themselves and against one another. The bloody and protracted conflicts generated by religious division eventually persuaded many that they could be brought to an end satisfactorily only by an agreement to live and let live. Toleration might be difficult for true believers but it was preferable to civil and international strife. Doubtless the course of European history has much to do with the close association of toleration with religion, but there is also a less contingent reason for that association.

To hold a religious belief is necessarily to dissent from and to reject other religious beliefs. To be a Christian is necessarily to reject the Islamic belief that Muhammad was God's Prophet and that the Koran is the word of God. To be a Muslim is to reject the Christian belief that Jesus Christ was God Incarnate. To be an Orthodox Jew is to reject the beliefs of both. The same is true for those subscribing to a particular variant of a faith. The Protestant and the Roman Catholic, insofar as their beliefs differ, reject one another's beliefs. The same holds for the Sunni and the Sh'ia, and for both in relation to the Ahmadi. Of course, there are usually overlaps of belief between different versions of the same faith and sometimes between different faiths. But the observations I make here are sufficiently true to sustain the simple point I intend them to make: the negativity or objection that is intrinsic to the idea of toleration is also intrinsic to differences of religious belief. That is why a live-and-let-live approach to religious differences typically entails toleration.

It is not always religious beliefs merely as such that provide occasions for toleration. More frequently it has been deeds: manifestations of belief through teaching and preaching or through belief-based practices such as forms of worship and forms of dress. But if those deeds have attracted the censure of those subscribing to other religions, it has usually been because the beliefs they manifest have been judged mistaken.

In characterising different religious beliefs as conflicting beliefs, I do not mean to imply that those conflicts must turn into physical conflicts or conflicts of other sorts. Arguments for toleration seek to persuade us that conflicts of belief need not and should not turn into other sorts of conflict. I make only the elementary point that different religious beliefs are rival religious beliefs and that those who hold a particular religious belief intelligibly disagree with and disapprove of the beliefs that are its rivals. A religious faith may itself require its adherents to be tolerant of other faiths, but that is entirely consistent with the requirement's being a requirement to 'tolerate', with all that that implies.

Sometimes we are encouraged to 'celebrate' rather than merely tolerate religious differences, but it is difficult to see how that injunction can be consistent with taking religious differences seriously. The non-believer who looks upon the religious world with benign incredulity may find much to celebrate in its diversity. Without Christianity we would not have Bach's B-Minor Mass and without Islam we would not have the splendid mosques of Cordoba and Samarkand. But the serious adherent of a religious faith cannot be similarly approving of faiths different from his or her own. For the devout Muslim, it must be a matter of profound regret that so many people fail to recognise Muhammad as God's Prophet and the Koran as the word of God. For the devout Christian, it must be singularly unfortunate that not everyone recognises that Jesus Christ was the Son of God and that the path to salvation lies in faith in him. How can either celebrate the existence of mistaken beliefs and the erroneous ways of life they inspire? The Abrahamic faiths are, of course, not typical of all religions, especially not of polytheistic religions, some of which can be more relaxed about, and more accommodating of, other faiths. But, insofar as different religions present us with different and rival bodies of belief, their adherents have ample reason to view negatively the faiths and the unbelief they reject. We can celebrate many sorts of difference, such as those of literature, music, art and cuisine, but not every sort of difference need be a matter for celebration.

The association of toleration with belief is not unique to religion. Political beliefs are less frequently discussed as objects of toleration, yet it is crucial to the functioning of democracy that people should tolerate the propagation of political opinions with which they disagree and tolerate challenges to political beliefs to which they passionately adhere. Toleration of different and conflicting intellectual claims is similarly crucial to the conduct of academic inquiry. Not every difference of belief will provide an occasion for toleration. Some beliefs concern trivial matters and some differences of non-trivial belief may evoke little more than a shrug of the shoulders. But the capacity of religious, political, and intellectual beliefs to evoke serious disagreement amongst their holders is too obvious to need argument. In contexts of serious disagreement, mutual toleration, with its attendant negativity, is not second-best; it is the best there can be.

John Rawls is a contemporary political philosopher who has taken that point very much to heart. He makes mutual toleration a feature of his just society. For Rawls (1993), the existence of different and conflicting religious beliefs is an entirely intelligible feature of the contemporary world. Religious differences are but one manifestation of the reasonable disagreement that we should expect to find amongst the citizens of a free society. Rather than seeking to reduce or eliminate those differences, a just society would seek to provide for them and to do so fairly. In particular, the citizens of a just liberal society would refrain from using political power of any kind either to promote their own religious beliefs and ways of life based upon them, or to disadvantage the different beliefs and ways of life of others. Toleration of that sort is, for Rawls, not a regrettable concession to human frailties but a welcome feature of a just society; it belongs to 'ideal theory'.

Rawls generalises that even-handed idea of toleration from the particular case of religion to 'comprehensive doctrines' generally. Those include philosophical and moral, as well as religious, doctrines. Such doctrines are subject to reasonable disagreement and a just society would distribute freedom and resources in a fashion that was neutral amongst its citizens as the adherents of different comprehensive doctrines.

So we have in Rawls an attempt to extend the idea and practice of toleration from the classic case of religious differences to the more general differences that characterise contemporary plural societies. Yet it is noteworthy that Rawls still characterises those differences as *doctrinal*. To that extent, the plurality for which his toleration provides is still a plurality of beliefs and of belief-based ways of life.

Toleration and identity

Are the salient differences of contemporary plural societies really doctrinal in nature? Those differences are now more commonly construed as differences of identity and many such differences have little or nothing to do with belief. Differences of race, nationality, gender, sexual orientation, or disability for example do not manifest differences of belief. Beliefs and doctrines may relate to those identities but the identities themselves are not doctrinal or belief-based. Some sources of identity remain significantly tied to belief, the most prominent example being religion. Belief and belief-based differences can also figure significantly in cultural identities, which are often closely intertwined with religious identities. But conceiving a religion or culture as a feature of a person's identity is significantly different from conceiving it as a body of belief to which a person subscribes. Most obviously, the person's religion or culture becomes part of who that person *is*, so that we cannot detach our treatment of a religion or culture from our treatment of the person whose religion or culture it is.

Identity is a varied and complex phenomenon (Appiah, 2005; Fukuyama, 2018; Parekh, 2008). I shall not try to do justice to that variety and complexity here. To get at the issue that is my concern, I want to focus on the logic of identity – what the idea of identity implies – and to consider how that logic bears on the appropriateness and acceptability of toleration as a response to identity. Would the members of a plural society ideally respond to one another's different identities with the same toleration that they would ideally show to one another's different beliefs?

There are a number of reasons why toleration, traditionally conceived, loses much of its appeal in the case of identities. One is that, whereas different beliefs or doctrines (addressed to the same issue) are intrinsically conflictual, different identities are not. My being of a particular race or ethnicity or gender or sexual orientation need not pit me against the different identities of others. So the disapproval or objection that is intrinsic to both the idea of toleration and conflicts of belief is not similarly intrinsic to differences of identity. The word 'difference' is sometimes intended to convey just that point: identity differences are *mere* differences, so that there is no obvious reason why they should attract disapproval, dislike, or rejection. We know that, as a matter of fact, differences in identity are sometimes caught up in vicious conflicts and I comment on that reality later. But, insofar as different identities do not 'disagree', they do not provide the same reason as different beliefs for mutual objection.

A second consideration takes us to the heart of identity. People's identities constitute who they are. To deprecate or be hostile to their identities is to deprecate or be hostile to their very selves. We are understandably reluctant to condone the indictment of people merely for being who they are. Yet, if we laud the toleration of different identities, we seem by implication to condone the negativity towards identities that toleration entails.

Does that have to be so? In the case of differences of belief, we can separate a belief from the person who holds it. We can therefore be dismissive of a belief without being dismissive of its holder. A prominent argument for toleration relies on just that thought. I may disapprove of your belief and the way of life it sanctions, but I am duty-bound to respect your personhood, which in turn means I must respect your right to hold your beliefs and to live in conformity with them (e.g. Forst, 2013). Can I not, in a similar fashion, take a negative view of your identity while respecting you as a person so that my negativity in no way diminishes your status as a person? That is a more difficult trick to perform. Suppose I say that I love animals but I hate mice. Can I simultaneously love the animality of mice but hate their miceness? It is hard to see how I can. What I am really saying is that I love animals with the exception of one type of animal – mice – which I hate. Thus, simply stated, I hate those animals that are mice. Analogously, suppose I say that I respect all persons but I despise Icelanders. Can I simultaneously respect the personhood of Icelanders, while despising them as Icelanders? Again, that is a difficult feat to pull off. What I am really saying is that I respect all persons with the exception of Icelandic persons whom I despise. So, again, simply stated, I despise those persons who are Icelandic.[1] Identities do not merely supervene on persons. They mark out different sorts or categories of person. That is why to despise or hate a particular identity is to despise or hate the person whose identity it is. Once we see the world in terms of identities, it becomes much more difficult to hate the 'sin' without also hating the 'sinner'. That gives us reason to question whether we are right to hate the 'sin' – the identity – at all.

That objection can be compounded by another. Some identities are self-assumed; for example, those who are Punks or Goths or Hipsters are so by choice. But the identities that have the greatest salience in contemporary societies tend to be ascriptive; they are identities people discover themselves to have, if only in the eyes of others, rather than identities they themselves assume. That is true of race, ethnicity, and culture. It is also true of gender; even those who change their gender typically conceive their transgendering not as a wilful act of self-invention but as a response to their discovering who they really are. Sexual orientation too is now generally understood to be a matter of discovery rather than choice. Religion is something of a special case in that people can forsake the faith of their family or community and adopt a religion that is no part of their heritage. Yet the great majority of the world's population do not. Their religion remains the religion into which they have been socialised so that it too functions, for themselves and for others, as an ascriptive identity.

Insofar as people's identities are ascriptive, we are likely to find it particularly unacceptable that others should view them negatively. For good reason, we frown upon disapproval or denigration of people for features of themselves for which they bear no responsibility. So the disapproval and dismissal that we can find entirely acceptable in respect of doctrinal differences, and which in turn can make toleration of those differences acceptable, can be entirely unacceptable in respect of different identities.

That leads us to a final dissonance between toleration and identity. Disapproval of an ascriptive identity makes little sense. It makes little sense to cast a negative moral judgement upon people for features of themselves over which they have no control. But if disapproval is off the agenda for identities, dislike is not and people's toleration can still be occasioned by their dislike. If they tolerate what they dislike, is that not something we should welcome? Not necessarily. While we can have good reason to accept people's dissent from the beliefs of others, we do not have similarly good reason to accept their disliking the identities of others. There is nothing reprehensible in disagreeing over beliefs or in responding to that disagreement with toleration. But there is something reprehensible in people's disliking or hating one another's identities and in their indulging their dislike or hatred by merely tolerating the objects of their dislike or hate. Rather than being tolerant of others, the dislikers and the haters should shed their dislike and hatred.

That, however, may be the counsel of perfection. One identity may define itself in opposition to another. That is often evident in cases of national identity. Even in those cases, hostility is not simply an artefact of identity. The identities at issue will have historical hinterlands which explain why their possessors have come to view one another negatively. But, insofar as the bearers of identities do conceive one another with hostility, toleration is very much to the point. We and they have good reason to prefer toleration to open conflict. For example, the issues and tensions that led to civil conflict in Northern Ireland are still very much features of the province, but most of its population prefer the current peace, to which mutual toleration is essential, to the 'Troubles' that preceded it. In some circumstances, then, we have good reason to welcome rather than to frown upon the toleration of identities. But we may still take the view that civil strife and bloodshed are pathological rather than normal or necessary features of identity difference, and so locate toleration, along with the circumstances that make it necessary, in the 'non-ideal' realm. As long as conflict persists, toleration may be the best we can hope for, but it is still second-best. When circumstances are as they should be, identities should, we may insist, make no call upon our toleration.[2]

Toleration and tolerance

Before settling for that simple picture, we should pause to consider whether negativity towards what we tolerate must be a feature of toleration. Several commentators suggest that it need not. They do not deny that negativity can be a feature of toleration; they deny only that it has to figure in every form of toleration. Toleration, they claim, should be understood more generously to include stances such as indifference or positive endorsement, which carry no implication of negativity (e.g . Apel, 1997; Creppel, 2003, pp. 3–4; Walzer, 1997, pp. 10–12). Peter Balint and Preston King distinguish between 'toleration' and 'tolerance', giving 'toleration' a broad meaning that encompasses any response to difference that is not intolerant and using 'tolerance' to describe one type of toleration – that in which the tolerator objects to but does not hinder that which he or she tolerates; i.e. the sense in which 'toleration' has been used in this chapter (Balint, 2017, pp. 5–6, 23–35; King, 1998, pp. xii–xv, 12–13).

What motivates this more generous usage? Sometimes it is the thought that what really matters about toleration is the absence of intolerance, so that any stance that is not intolerant, including indifference and positive affirmation, warrants inclusion in toleration. Sometimes it derives from the conjunction of toleration and pluralism: toleration aims to accommodate pluralism, so any position that contributes to that accommodation, whether it be merely putting up with pluralism or embracing it with enthusiasm, can be accounted tolerant.

If we can redefine 'toleration' so that it becomes possible to tolerate even though we neither disapprove nor dislike, we can tolerate identities without running into any of the objections catalogued in the previous section. Enlarging the meaning of 'toleration' in that fashion does not dispose of the issue I mean to address in this chapter: is the toleration that is appropriate to differences of belief similarly appropriate to differences of identity? Moreover, we might object that removing negativity from the meaning of 'toleration' diminishes its descriptive utility: if it ceases to distinguish 'objecting but enduring' from indifference or positive affirmation, it no longer picks out a stance that differs importantly from them. On the other hand, language has a life of its own and one use of the word 'tolerant', which pushes beyond the idea of 'putting up with', has gained widespread currency and is significant for what people may mean when they commend tolerating identities.

Suppose that someone adopts a studied commitment to not viewing human diversity negatively. While those around her respond with suspicion and hostility to the new, the unfamiliar and the alien, she does not. When she confronts a way of life different from her own, she views it with interest, empathy and an open mind. If she finds it prima facie odd and in conflict with her own preconceptions, she looks at it more closely, on the working assumption that its apparent oddness and dissonance is more likely to derive from her own lack of understanding than from something defective in the way of life itself. Thus, far from taking exception to the new, the unfamiliar, and the different, she refrains from doing so. How should we describe such a person?

It would be odd to deny her the description 'tolerant' and she deserves to be described in that way precisely because she refrains from disapproval and dislike. It is her open-minded and generous approach to difference that marks her out as the very model of a tolerant person.[3] Since the word 'tolerance' exists, I shall use it to badge this sort of 'being tolerant' and to distinguish it from toleration traditionally conceived, though my giving 'tolerance' that meaning is little more than stipulative.[4] When people hold out the ideal of a tolerant society, it is frequently tolerance in this sense that they mean to commend. They conceive a tolerant society not as one whose members object to but endure one another's differences but rather as one whose members are at ease with one another's differences because they take no exception to them. For several years, the tag-line of the British Home Office was "building a safe, just and tolerant society". That slogan was no doubt the product of a public relations exercise rather than deep reflection on the human condition, but it is most unlikely that the Home Office aspired to build only a society whose members disliked but endured one another. Rather, they aimed for a society that was harmonious and at peace with itself because its members happily accepted one another's differences.

Is there any connection between tolerance so understood and toleration traditionally conceived? One is that both entail forbearance: in one case the tolerator forbears from impeding or hindering, while in the other he or she forbears from disapproving or disliking. It is not difficult to understand how the first usage has given rise to the second. Another is that we are inclined to speak of tolerance only in respect of differences to which others do not, or have not, or might not, show tolerance or toleration. We need not dwell on these semantic questions. What matters is that this understanding of 'being tolerant' is now current. We should be careful therefore not to mistake what people mean when they commend us to 'tolerate' one another's differences, and not to chastise them unjustifiably for moral error or moral inadequacy in commending tolerance of differences.

If tolerating identities takes the form of tolerance, might it claim a place in ideal theory? For some, it may still fall short. Mere acceptance is not enough; differences should be met with positive endorsement, celebration, and applause. But that view is rather odd. Differences are just the way things are; there is no reason why they should always occasion paroxysms of delight. We may, of course, find reason to admire and praise an identity which differs from our own; we may, for example, find much to admire in a culture that affords the identity of another. But, if we do, we are in the business of according appraisal respect rather than recognition respect (Darwall, 1977). Appraisal respect by its very nature must wait upon, and depend upon, the findings of an appraisal and we cannot assume those findings will justify equal respect for all. Appraisal, if it is meaningful and honestly conducted, is most unlikely to issue in equal praise all round. If we demand equal respect for identities and make that demand a matter of moral obligation, our demand makes sense only as one for recognition respect, which entails ascribing equal status to the recognised and treating them accordingly.

My account of differences of belief and identity and of how they relate to toleration and tolerance has been simple and stylised, though not, I hope, too simple and stylised to relate to the way things are. I now want to comment on two issues which illustrate how claims of belief and identity can vie with one another in the public domain: (1) the extent to which people should be free to criticise and satirise the religious beliefs of others and (2) the extent to which a society should accommodate the demands of particular religious faiths when they conflict with the society's rules and practices. In considering these issues, I shall focus on toleration rather than tolerance.

Belief, identity, and challenges to religion

By 'challenging' a religion, I mean criticising or satirising it in ways that upset a substantial number of its adherents. Challenges of that kind have given rise to a number of well-known episodes, such as the Rushdie Affair which was sparked by the publication of Salman Rushdie's *Satanic Verses* in 1988 and which persisted for a decade thereafter. More recently, the publication of cartoons of Muhammad in the Danish newspaper, *Jyllens-Posten*, in 2005 caused outrage amongst Muslims, not only in Denmark but across the world. Martin Scorsese's, *The Last Temptation of Christ*, first screened in 1988, and *Jerry Springer the Opera*, first staged in 2003 and broadcast in the United Kingdom by the BBC in 2005, provoked protests from Christians, if of a more restrained sort. Here I do not wrestle with the complexities of those cases. I consider only how toleration bears on them and what hangs on whether we conceive them as challenges to belief or identity.

Suppose we regard a challenge to a religion as a challenge to its beliefs. For whom does that challenge raise issues of toleration? Ultimately, the answer is all members of the society in which it arises and potentially, given the ease with which these issues now become global, the world at large. But more immediately they tax the toleration of those whose beliefs are subjected to criticism or satire. Toleration would not require the believers to endure objectionable treatment of their faith without answer or protest, but it would require them not to press for legal prohibition of that treatment or to foster a public culture that deems it intolerable. Might the challengers themselves be charged with intolerance? They might be adjudged not to show tolerance, as I have defined that term, but it is more difficult to sustain the charge that they fail to show toleration, given that criticism and satire neither remove nor diminish the freedom of believers to hold, profess, and practise their faith. Challengers certainly behave differently from the way believers would have them behave, but toleration becomes absurd if it requires us to behave only as others wish (Jones, 2018, pp. 30–34).

If we stay within the realm of belief, the case for tolerating critical scrutiny and satirical comment is easily made. The whole panoply of arguments for free expression can be mobilised in defence of that scrutiny and comment. Those arguments apply with especial force to religions which, like Christianity and Islam, are evangelical in mission and proclaim truths that are ostensibly universal in relevance. Their adherents cannot plausibly surround their faith with 'keep out' notices and insist that it is properly of concern to none but themselves. That is not to say that, within the realm of belief, everything must go the way of free expression. The beliefs that are subject to criticism and satire are typically beliefs that matter to their bearers because they concern sacred subjects, whose obscene, scatological, mocking, or disrespectful treatment amount to acts of desecration. Challengers can reasonably be expected to be mindful of the meaning their deeds have for those whose beliefs they target. Moreover, adverse treatments of religion are sometimes merely gratuitous taunts that find little defence in the standard arguments for free expression. But objecting to criticism and satire because they treat sacred subjects wrongly, thoughtlessly, or tastelessly is quite different from objecting because they assault the identity of believers.

When we shift perspective from belief to identity, the nature of the challenge becomes very different. If we conceive being Christian or Muslim as a matter of identity, we sink Christianity or Islam into the being of the person who bears it. Thus, in attacking a belief, we assault its bearer and, in ridiculing a belief, we humiliate those whose belief it is. The party wronged by the Danish cartoons ceases to be Muhammad and becomes those who identify as Muslim, just as the party wronged by *Jerry Springer* ceases to be Jesus Christ and becomes those who identify as Christian. If that is how we should understand what is going on, it is more difficult to see why Muslims and Christians should tolerate their challengers. The attackers and mockers attack and mock not a set of epistemic propositions but a set of people and those people can reasonably find that treatment intolerable. If expressions harm people, the case for their being tolerated falls away. Moreover, charging the challengers with intolerance now seems mistaken in a quite different way: the charge is inappropriate not because the challengers are innocent but because it indicts their conduct in too feeble a way. The identity perspective effects another change: when we translate a faith into an identity, we make plausible a conception of the faith as the private possession of those who embrace it. A faith becomes the property of those whose faith it is and a property upon which outsiders should not trespass. It also becomes easier to recast criticism and satire as manifestations of hatred or dislike, since, from the perspective of identity, critics, and satirists target not beliefs but people. It is in this context that the notion of Islamophobia makes most sense, since a 'phobia' describes an aversion or revulsion rather than doubt, disagreement, or dissent. Thus, the nature of what is at issue in challenging a religion is transformed if we conceive it as challenging an identity rather than a belief.

There is something of a paradox in believers, as distinct from others who may speak up for them, appealing to the claims of identity. Theologically, the Danish cartoons were wrong because of the way they treated Muhammad and *Jerry Springer* was wrong because of the way it treated Jesus Christ. Thus, theologically, the Muslim or the Christian who protests, '*I* am the wronged party; the challengers traduced my *identity*', puts the emphasis in the wrong place. But people's theology does not always govern their psychology. Moreover, there is no point in believers' protesting in theological terms to those who do not share their theology. They can mobilise claims of identity as claims that should be acknowledged by others who do not share their faith, even if they themselves do not value their faith principally as an identity.

Here I shall not try to arbitrate between the perspectives of belief and identity. I note only that the difference between them helps explain why people see what is at issue in challenges to religion so very differently and why so much of the debate that the challenges stimulate seems to be at cross-purpose. It is not plausible to dismiss either perspective entirely but neither is it easy to see how the conflicting claims they generate can be harmonised.

Belief, identity, and accommodation

By religious accommodation I mean a society's making provision for the demands of a particular faith. A society's rules or policies sometimes conflict with the practices of a religion such that they disadvantage or 'burden' its members. Through accommodation a society seeks to relieve those burdens. Examples are measures, to be found in some European societies but not all, that exempt turban-wearing Sikhs from the legal obligation to wear a safety helmet if they ride a motorcycle or work on a construction site, and Jews and Muslims from animal welfare legislation requiring that animals be stunned before slaughter. The issue of accommodation also arises in human rights law, when courts are tasked with defining what precisely the right to manifest religion or belief, under Article 9 of the European Convention (ECHR), is an entitlement to; and it features in law governing indirect religious discrimination, which requires employers and providers of goods and services to exempt religious adherents from rules or practices if not doing so would disadvantage them relative to the adherents of other faiths or none.

How does religious accommodation relate to toleration? Accommodation is simply part of the general pattern of arrangements through which a society provides for the religious freedom to which the society is committed. But is not accommodation a more specific exercise in toleration than that? It is, but describing that more specific exercise as an exercise in religious toleration can mislead. What is tolerated is not so much the particular religion at issue as the exemption of its adherents from a rule with which others have to comply. For example, the considerations that argue for the compulsory wearing of safety helmets apply as much to Sikhs as they do to other motorcyclists or construction workers, and those that argue for animals to be stunned before slaughter apply as much to animals slaughtered by Jews and Muslims as they do to animals slaughtered by others. Hence, the reasons for having the laws are as relevant to the adherents of those religions as they are to other members of the society. But it does not follow that granting an exemption to those adherents must be done grudgingly or with misgiving. On the contrary, when a society grants an exemption, it can be understood to recognise that a particular religious group has good reason to be exempt – a reason not possessed by others – and that, all things considered, exempting the group is the right thing to do. Toleration does not cease to be toleration simply because it is justified.

That said, an exemption remains an exceptional measure. What can justify it? Belief provides a more convincing case than identity. If, for example, the issue of exemption arises because an employer's dress code clashes with that of a religion, we need to explain why religious employees should be eligible for an exemption while fashionistas should not. Invoking identity is unlikely to do the trick. Why should the wish of the religious to express their identity count for more than the equivalent wish of fashionistas? It is religious employees' believing themselves to be subject to an obligation or to some other imperative of their faith that most convincingly distinguishes their claim from the fashionistas'. Similarly, if the Sikh registers a claim to carry a kirpan knife even though law prohibits knife-carrying in public, or the Roman Catholic doctor or nurse wishes to be exempt from participating in abortions, or the Muslim wants time off to attend Friday prayers, or the Orthodox Jew or the Sabbatarian Christian requests not to work on their faith's day of rest, the most obvious reason we have to take those claims seriously resides in the conscientious beliefs from which they derive. We would be justifiably unimpressed if each sought to vindicate his or her claim merely by declaring, 'it is part of my identity'.

Belief, then, provides a more compelling case for religious accommodation than does identity. However, the issue of accommodation differs in character from that of challenges to religious faith, considered in the previous section. In the case of challenges to a faith, we saw that the logic of belief could pull in a *different* direction from the logic of identity. Insofar as the claims of religious identity bear on the case for religious accommodation, they will push in the *same* direction as those of belief. Even so conflicts between belief and identity can arise in policy on accommodation but this time as conflicts between the claims of religious belief and those of non-religious identity. That conflict has been most conspicuous in the area of discrimination law; I shall consider it through the case of British law on discrimination.

The UK Equality Act 2010 prohibits direct and indirect discrimination in employment and in the provision of goods and services in respect of a number of 'protected characteristics'. Those characteristics include race, gender, sexual orientation, gender reassignment, marital status, disability, and age. They also include 'religion or belief'. However, as well as being a protected characteristic, religious belief figures in the Act as a claimant that competes with other protected characteristics. That competition arises because some religions want, for doctrinal reasons, to discriminate, on grounds such as sexual orientation, in whom they employ and to whom they provide goods and services. How does the Act provide for that competition?

It allows 'organised religions' (churches, mosques, synagogues, etc.) to discriminate in employment not only on grounds of religion or belief but also on grounds of gender, gender reassignment, sexual orientation, and marital status, provided that that discrimination is required for compliance with the religion's doctrines (the 'compliance principle') or to avoid conflict with the strongly held convictions of a significant number of the religion's followers (the 'non-conflict principle').[5] It also allows 'organisations relating to religion or belief' to discriminate on grounds of sexual orientation in the provision of several types of good or service, provided again that that discrimination is justified by the compliance principle or the non-conflict principle.[6]

The Act's allowing discrimination only insofar as it is warranted by the compliance principle or the non-conflict principle indicates that it exempts organised religions out of deference to their beliefs rather than merely their identities. Without those exemptions, discrimination law could be at odds with the doctrines and practices of organised religions, which might in turn violate their right to freedom of religion under Article 9 of the ECHR.

Setting aside that legal matter, should we regard the exemptions granted to organised religions as exercises in toleration? From the perspective of British public policy, arguably we should. That policy deems it unjust that citizens should have unequal opportunities in employment or in their access to goods and services in respect of any of the characteristics recognised in the Equality Act. More generally, it is committed to according equal status and equal recognition to its citizens in respect of those characteristics. Thus, in exempting organised religions from discrimination law, public policy is subordinating its own commitments to equality of opportunity and equality of status to the doctrinal commitments of organised religions. That is not to say that public policy is at odds with itself and allows something that, based on its own commitments, it should not. The all-things-considered position of public policy must include a judgment on the respect it owes people as religious believers.[7] As we observed earlier, toleration does not cease to be toleration simply because, all things considered, it is the right or best policy.

The Equality Act is, however, far from generous in the toleration it extends to religious belief. First, it narrowly circumscribes the range of posts in respect of which organised religions may discriminate on otherwise protected grounds.[8] Second, it does not extend to individual believers the exemptions it grants to organised religions. Religious individuals, as employers or providers of goods and services, may not discriminate on grounds of protected characteristics, even though their religious beliefs might sanction that discrimination. Why not? The most plausible justification lies in the greater 'immediacy' with which organised religions would be disadvantaged if there were no exemption: they could find themselves legally obliged to defy their own doctrines. By contrast, if a religious individual who occupies a role outside of an organised religion finds that he or she cannot, for religious reasons, discharge that role consistently with discrimination law, he or she can escape his or her dilemma by taking up a different role.

That is not to say that, with respect to discrimination law, the claims of religious belief to toleration dovetail neatly with those of identities to recognition and protection. On the contrary, the two claims clearly compete and the Equality Act deals with their competition by compromising between them. Whether a plural society should use accommodation to enlarge the space of toleration afforded to religious groups remains deeply controversial, but it is especially controversial when that accommodation is secured at the expense of a non-religious identity.[9]

Conclusion

We began by asking how much space there should be for toleration in a plural society ideally conceived. The answer, we have seen, depends significantly on the sort of pluralism a society exhibits. Insofar as its pluralism consists in a plurality of beliefs, a society will normally have ample reason to embrace toleration as an ideal and to embed that ideal in its arrangements and its public culture, even though negativity remains a feature of toleration. Insofar as its pluralism consists in a plurality of identities, it has much less reason to view toleration favourably, precisely because toleration harbours negativity. For a society in which identity differences are sources of hostility and conflict, mutual toleration amongst its citizens may well be a condition to which it should aspire; but toleration will then be an instrument for coping with the non-ideal circumstances of hostility and conflict, rather than the condition that would ideally prevail in a society characterised by identity differences. Ideally, the relations prevailing among citizens bearing different identities would be not fall below those of mutual tolerance (as I have defined 'tolerance').

Dealing with pluralism is not simply a matter of disentangling beliefs from identities so that we can respond to each appropriately. Religions are both bodies of belief and sources of identity. As we have seen, what is at issue in 'challenges' to religion, and how we think a society should respond to them, will depend on whether we conceive them as critiques aimed at beliefs or assaults targeted at identities. In addition, the space of toleration demanded for religious belief can compete with the demands of non-religious identities for respect and recognition, as is often apparent when efforts are made to accommodate religious belief. Thus, despite the favourable aura that still tends to surround the idea of 'being tolerant', the task of determining the space toleration should occupy in a plural society is not an easy one, nor can it easily escape controversy.[10]

Notes

1 I have nothing against Icelanders. They strike me as a group who are most unlikely to suffer from this negative appraisal and who are therefore, I hope, unlikely to be upset by my using them in this example.

2 All of these objections to tolerating identities would disappear if the idea of toleration could be recrafted to incorporate the recognition of identities. For an imaginative and sustained effort to perform that task, see Galeotti (2002). For a sceptical response, see Jones (2018, pp. 129–152).

3 Cf. John Horton (1996, p. 38): "The tolerant person is not a narrow-minded bigot who shows restraint; he or she is not someone with a vast array of prejudices about others' conduct but who nonetheless heroically restrains him- or herself from acting restrictively toward them. The restraint involved in toleration is not exclusively of action but also of judgment. The tolerant person is not too judgmental toward others. In becoming less judgmental, a person becomes more tolerant". Similarly Glen Newey (2013, pp. 6–7): "the term 'tolerant' can be applied to people who are not disposed to feel disapproval of others. For instance, the Netherlands has often been described as a tolerant society, which is usually taken to mean not that the Dutch habitually feel

strong disapproval which they strenuously suppress, but that they readily accept the differences of others".

4 My use of 'tolerance' is therefore quite different from the meaning it has for Peter Balint and Preston King, which I noticed earlier.

5 Equality Act 2010, schedule 9, para 2.

6 Ibid., schedule 23, para 2.

7 The way in which many organised religions currently discriminate on grounds other than religion or belief is controversial, inside as well as outside those religions. But we should distinguish between two questions: (1) is a religion justified in discriminating with respect to characteristics other than religion or belief? (2) if it is not, should the state intervene to prevent that discrimination? Those who answer 'no' to question (1) need not answer 'yes' to question (2). They may believe that, while a doctrine or practice of an organised religion is mistaken, it is for the religion itself to correct that mistake rather than for the state to usurp the religion's right of self-governance.

8 Equality Act 2010: Explanatory Notes, paras 799–800.

9 For discussion of some high-profile discrimination cases in which the claims of religious belief have been pitted against those of identity, particularly sexual orientation, see Jones (2018, pp. 69–92), and Malik (2013).

10 I am grateful to Luiza Bialasiewicz and Valentina Gentile for their comments on an earlier version of this chapter.

References

Addis, A. (1996). On human diversity and the limits of toleration. In: I. Shapiro and W. Kymlicka, eds., *Ethnicity and group rights*. New York: New York University Press, pp. 112–153.

Apel, K.O. (1997). Plurality of the good? The problem of affirmative tolerance in a multicultural society from an ethical point of view. *Ratio Juris*, 10(2), pp. 199–212.

Appiah, K.A. (2005). *The ethics of identity*. Princeton, NJ: Princeton University Press.

Balint, P. (2017). *Respecting toleration*. Oxford: Oxford University Press.

Beneke, C. (2006). *Beyond toleration: The religious origins of American pluralism*. Oxford: Oxford University Press.

Brown, W. (2006). *Regulating aversion: Tolerance in the age of identity and empire*. Princeton, NJ: Princeton University Press.

Brown, W. and Forst, R. (2014). *The power of tolerance: A debate* (L. Di Blasi and C. Holzhey, eds.). New York: Columbia University Press.

Creppel, I. (2003). *Toleration and identity*. New York: Routledge.

Darwall, S. (1977). Two kinds of respect. *Ethics*, 88(1), pp. 36–49.

Forst, R. (2013). *Toleration in conflict*. Cambridge: Cambridge University Press.

Fukuyama, F. (2018). *Identity*, London: Profile Books.

Galeotti, A.E. (2002). *Toleration as recognition*. Cambridge: Cambridge University Press.

Heyd, D. (2008). Is toleration a political virtue? In: M. Williams and J. Waldron, eds., *Toleration and its limits, nomos XLVIII*. New York: New York University Press, pp. 171–194.

Horton, J. (1996). Toleration as a virtue. In: D. Heyd, ed., *Toleration: An elusive virtue*. Princeton, NJ: Princeton University Press, pp. 28–43.

Jones, P. (2018). *Essays on toleration*. London: ECPR Press/Rowman & Littlefield.

King, P. (1998). *Toleration*. 2nd ed. London: Frank Cass.

Malik, M. (2013). Religion and sexual orientation: Conflict or cohesion? In: G. D'Costa, M. Evans, T. Modood, and J. Rivers, eds., *Religion in a liberal state*. Cambridge: Cambridge University Press, pp. 67–92.

Newey, G. (1999). *Virtue, reason and toleration*. Edinburgh: Edinburgh University Press.

Newey, G. (2013). *Toleration in political conflict*. Cambridge: Cambridge University Press.

Parekh, B. (2008). *The new politics of identity*. Basingstoke: Palgrave Macmillan.

Rawls, J. (1993). *Political liberalism*. New York: Columbia University Press.

Walzer, M. (1997). *On toleration*. New Haven: Yale University Press.

4 Infrastructures for living with difference

Dan Swanton

I begin with a trouble. My trouble is with the idea of tolerance. My concern in this chapter is to trouble talk of tolerance in the context of urban multiculture in British cities. Tolerance is routinely deployed in acts of public and political speech – as an adjunct to smiling (or gritted-teeth) rhetoric of inclusion; as a shifty British value of live and let live that at once denies and reproduces racism (Lentin and Titley, 2011); or as an explanation for the surfacing of cultural or religious practices that are deemed unacceptable by some group or another. These different mobilisations of tolerance in the context of living with difference introduce the trickiness and ambivalence of tolerance. I am uncomfortable with the idea of tolerance, and the ways it becomes entangled with diagnosing, managing, and imagining social relations in multicultural places.

This trouble spurs me to thinking differently about spaces of tolerance in cities. In the chapter I side-step the established conceptual terrains of tolerance, where tolerance is understood either as a political ideal and norm that forms a baseline for dissensual political in models of deliberative democracy (Gill, 2012; Jones, 2018; see also chapters in this volume by Jones, Balint, Laegaard, and Gentile), or it is treated as a discourse of power and mode of governmentality (Hage, 1998; Brown, 2006). In my discussion, I push back on some of the more abstract discussions of tolerance in political theory, approaching the concept through an attention to the ethnographic details of multicultural places, and the messy lived realities of everyday spaces of tolerance. I attend to some of the everyday ways in which tolerance – alongside other affective orientations to difference – surface in the 'midst of the racism's ruins' (Back and Sinha, 2016) in British cities. Back and Sinha (2016, p. 521) are interested in documenting a "situated ethics of conviviality" and cultivating "a 'way of seeing' that is attentive to forms of division and racism alongside and sometime within 'multicultural convivialities". In the chapter, I play on this metaphor of ruins, and more specifically the possibilities of life in ruins (cf. Tsing, 2016). Ruins offer a segue into 'broken world thinking' that "asks what happens when we take erosion, breakdown, and decay, rather than novelty, growth and progress as our starting points" (Jackson, 2014, p. 221). If we shift our orientation to the world as one that is "almost-always-falling-apart", we become attuned to the processes of holding together that involve repair, improvisation, invention, fixing, and reassembling. I adopt a similar orientation to spaces of tolerance, examining how social norms and entrenched habits of (racialised) sense-making are negotiated, suspended or reworked.

In particular, I focus on two examples. The first works through the details of a gathering scene. It focuses on a very ordinary space outside a mosque in Edinburgh, where *something happened* after the terror attacks on two mosques in Christchurch, New Zealand on March 15, 2019. Examining a makeshift memorial to the victims of a suspected Far-Right terrorist, in this example I argue for the need to stay with the messy, the messed up, and the difficult. I read this fleeting memorial as a materialisation of repair work that points to new ways of being with Others and weaves new connections as a world falls apart. And so, in the midst of racism ruins and wreckage of state-sponsored multiculturalism and of social democracy, possibilities of life emerge as spaces of tolerance and conviviality are made. My second example draws on my recent research (Swanton, 2018) that develops writing on infrastructures, and how they might offer ways of recognising and crafting spaces for living with difference that use displacement and disruption to unlearn habits of race thinking, reboot forms of sociality, and reimagine forms of affective continuity that involve new attunements to being with Others. Focussing on photographer Mahtab Hussain's project You Get Me? I examine how his portraits craft encounters with young British Muslim men in ways that call out racist visual cultures fed by negative media coverage (Swanton, 2018). These portraits stage performative encounters that have the potential, I argue, to disrupt practices of looking and sorting that are entrenched in habits of race thinking; and they offer the ground for imagining new patterns, habits and norms of sociality. My argument is that projects like You Get Me? – alongside recognising the ongoing social projects that involve forms of conviviality, and other experiments, interventions, and practices – contribute to an infrastructure of living with difference that dwell on the concrete acts and contexts of social collaboration (Simone, 2004) to disrupt racisms, produce engaged publics, and nurture the situated ethics of conviviality (Back and Sinha, 2016)

Troubling tolerance

"But just because we agree that intolerance is bad, doesn't mean tolerance is good".

A little over ten years ago, Wendy Brown posed this question: "How did tolerance become a beacon of multicultural justice and civic peace at the turn of the 21st Century?" (2006, p. 1). Writing in the context of the United States, Brown is concerned with the depoliticising effects of the renaissance of tolerance talk in public and political discourse. Her worry is that, in a generation, tolerance had gone from being a code word for mannered racialism to a norm that was being uncritically promoted as something that is as fundamental to universal human dignity as the freedoms of speech and conscience.

Today, in the United Kingdom it feels like the political and public enthusiasm for tolerance heralded by the development of policies of multiculturalism is fraying. In what Lauren Berlant (2011) might describe as the 'glitch' of the present moment – that includes states of emergency fuelled by economic crises, 'migration crises', and terror attacks – we witness popular backlashes against tolerance. Brexit, the election of Donald Trump, and a growing enthusiasm for far-right political organisations across Europe and beyond can be linked to recent histories of tolerance talk in liberal democracies that have been exploited to argue that 'we' have been tolerating the intolerable. These arguments follow the refrain that 'we' – always assuming the entitled position of host and custodian of the nation (Hage, 1998) – have tolerated too much migration, extreme religious ideologies, cultural difference, patriarchy or misogynistic practices, and so the list goes on. And yet, the idea(l) of tolerance in liberal democracy remains fundamental for governing diverse populations, responding to forms of extremism or justifying an ethics of care extended to displaced peoples.

My trouble with tolerance is that is it tricky. It is ambivalent. I find myself caught between at least two orientations to tolerance that rub along uncomfortably. These two positions are perhaps best embodied by Rainer Forst's writing on tolerance as an ethical practice and Wendy Brown and her work on tolerance as a discourse of power (Brown, 2006; Brown and Forst, 2014). These positions produce friction and I end up getting stuck. In what follows, I outline these key orientations to the idea of difference, before working towards an argument for infrastructures for living with difference.

Tolerance as a minimum for dissensual politics?

While recognising that tolerance is implicated in the flawed and damaging arrangements of power, tolerance can be viewed as a least bad option. Tolerance operates as a political ideal, a norm and a substantive common ground in contemporary liberal democracies, where a commitment to toleration is manifest in the design and operation of institutions (Jones, 2007, this volume; Balint, this volume; Gentile, this volume). In the context of religious practice, the ideal of tolerance shapes the rules and institutions within political orders such that various accommodations and exemptions can be justified through 'ideas of reasonableness and public reason' (Gentile, this volume) to protect freedoms of conscience, thought and speech, and to address religious discrimination (Jones, 2007). In liberal theory, tolerance becomes a political ideal that is examined in the context of accommodations, exemptions and public controversies where rival sets of beliefs that engender some form of disagreement or rejection are negotiated and settled (Jones, 2007, 2018). The commentary by political theorists on the ideal or norm of tolerance has been influential, but I would argue that it is partial. Tolerance is narrowly conceived and abstracted from the lived experiences of most people. The focus in political theory emphasises the negotiation of tolerance through related concepts of equality, public reason, reasonableness, justification, and legitimacy, drawing on empirical examples that tend to be restricted to episodes of public conflict (like the Rushdie Affair or Danish political cartoons), or exemplary legal cases that make judgements on complaints of religious discrimination (Jones, 2012; Gentile, this volume). The questions that animate my interests in tolerance are, however, more mundane and everyday. I am concerned with tolerance as an ordinary practice. In this sense, we might respond to the need for what Clive Barnett (2008, 2012) calls a "phenomenological approach to the political". That is to say a sense of the political that is orientated to the pragmatic, applied dilemmas of the everyday. In this vein, Rainer Forst argues for tolerance that exhibits "a sense of politics as practically-orientated action in the face of everyday dilemmas, tensions and conflicts" (Gill, Johnstone and Williams, 2012, p. 516 emphasis in original). Here, tolerance is an individual virtue and ethical practice that is a regular part of day-to-day life. It is concerned with the art and labour of getting along between family members, neighbours, colleagues. Tolerance is framed as a baseline for getting along based on practices of democratic justification. Either you accept a practice you object to if you do not have sufficient reason to reject that practice, or you find proper reasons to present to others when you think they should conform to a norm they do not agree with. What emerges is a 'grudging, gritted-teeth politics' (Modood and Dobbernack, 2011) imagined as a basis for inclusive, and yet dissensual, politics within liberal democracies (Gill, Johnstone and Williams, 2012).

My concern is that for tolerance to work as an individual and ethical practice in figurations of dissensual politics then the complexities, messiness, and difficulties of everyday lives need to be suspended. But tolerance does not only operate in a deliberative sense but also in a moral affective sense. We might think of the ways in which tolerance as dispersed and practised across these two senses – deliberative and moral. As Elizabeth Povinelli has argued (2011, p. 86), tolerance is located in the gut and the mind. And there is an undecidability about the location of tolerance. Where we register a gut feeling, we never quite know if this feeling had been camouflaged by the mind. Similarly, we cannot know whether the mind might merely mask a registering of affective forces that take over us (Povinelli, 2011). In the heat of cultural conflict, Forst's arguments about tolerance seems to require an artificial suspension of the world so that the ideas, ideals, and practices of deliberation, justification, adjudication, and reasoning can play out. These seem to be practices that are better suited to law courts, university seminar rooms, or opinion editorials in broadsheet newspapers, and not the places where day-to-day life is lived out. Indeed, it is perhaps not surprising that the examples that animate discussions of tolerance in liberal theory tend to focus on court cases that examine the public justification and negotiation religious protections or public controversies around freedom of expression (Jones, 2015, this volume; Gentile, this volume). But these ideals of tolerance translate awkwardly to the spaces of everyday life. The practically orientated action imagined in liberal theory seems to rely on the rationalisation or suppression of feelings and the complexities of lived spaces of tolerance, rather than a situated and negotiated ethics of conviviality. In other words, these appeals to tolerance fail to fully appreciate how "adjudication arises not via some transcendental gesture, no matter how fragile, but from a continual reflexively practised dwelling within the worlds to be adjudicated" (Povinelli, 2011, p. 33). Adjudication always takes place in the "queasy space of dwelling within these potential worlds" (ibid.).

A further concern with Forst's arguments for theorising tolerance as an individual virtue and ethical practice arises in cases of cultural conflict where the problems of social difference need to be resolved through public reason. In some cases public reason "must draw red lines across which difference cannot process, or a bracket must be put around the difference so that it can be removed from public debate until that time its challenge can be managed" (Povinelli, 2011, p. 31). Claire Jean Kim (2015, p. 6) illustrates this powerfully in her analysis of 'impassioned disputes' between multiculturalism and animal rights in the United States, where the cultural practices of nonwhites remain "hotly contested, and open ethnocentrism and intolerance persist". Through tales of animal use that include live animal markets in San Francisco's Chinatown, whale hunting by the Makah, and a charge against an NFL superstar, Michael Vick, for dogfighting, Kim highlights how multiculturalism and tolerance is far from settled. These 'impassioned disputes' provide a window on "contested and contradictory imaginings about struggles over race, species, nature and culture" and how to think through ethically and political competing moral claims (Kim, 2015, p. 8). In these disputes public discourse quickly locks onto 'internal pathologies', in ways that displace attention from historical and contemporary institutional arrangements and radicalised exclusion. Kim's (2015, p. 8) analysis of moments where there is a "disharmony of interests" highlights what Povinelli (2011, p. 31) would call an "internal incommensurability of late liberalism". These moments show how internal contradictions within deliberative democracy are projected onto the Other, "who are then called to present difference in a form that feels like difference but doesn't not permit any real difference to confront the normative world" (Povinelli, 2011, p. 31).

Tolerance as discourse of power and technology of governmentality

These concerns are powerfully taken up in by Wendy Brown (2006) in *Regulating Aversion*. She sees tolerance as an instrument of liberal governance and a discourse of power that legitimates white supremacy and state violence. Brown troubles how tolerance ontologises difference and conflict. She unpicks how tolerance sustains a "status of outsiderness for those it manages by incorporating; it even sustains them as a potential danger to the civic or political body" (Brown, 2006, p. 28). As such, discourses of tolerance manufacture unequal positions between tolerating and tolerated; they stratify and abject certain people from rights and equality without disturbing the norms of whiteness that marginalise them in the first place. Tolerance, then, is part of civilisational discourse that is cleansed of histories of slavery, colonialisms and imperialism, of fascism. Tolerance is an instrument of biopower that masquerades as neighbourly values and "cultural inheritance of 'live and let live' attitudes" (Brown, 2006, p. 38).

In recent years terror attacks on September 11, 2001 or July 7, 2007 have sparked a 'trembling of recognition' (Povinelli, 2011) and a crisis in Western liberal principles of tolerance. In the wake of these attacks, cultural difference takes on a new force. Povinelli (2011) likens these moments to shifting tectonic plates, where established orders are shaken and the shape and borders of social worlds are deformed and reformed. After 9/11 and 7/7, reactionary positions to cultural difference have gained momentum, spawning new and intensified regimes for the governance of social difference witnessed in declarations that multiculturalism has failed, and the need for a muscular deference of Western liberal principles. We are increasingly told that recognition has its proper limits, and that liberals always seek to overstep these limits because of their commitment to tolerance – something that Sune Laegaard's chapter examines in his discussion of the risks of securitising religion under militant liberalism. The violence of terror attacks is mobilised to reassert limits to recognition, and to not tolerate the intolerable, redrawing lines of tolerance in ways that enhance the life of some at the expense of others.

The corrosive effects of tolerance as a discourse of power in Britain are captured evocatively in Nikesh Shukla's (2016) edited collection of essays *The Good Immigrant*. The collection is fashioned as a response to the question: "What's it like to live in a country that doesn't trust you and doesn't want you unless you win an Olympic gold medal or a national baking competition?" (Shukla, 2016, no page). The 21 essays by writers of colour challenge how whiteness is reproduced as a social and cultural norm in Britain. Against wider norms that assume a universal experience that is white, the essays gather different kinds of universal experiences for people of colour in Britain: "feelings of anger, displacement, defensiveness, curiosity, absurdity – we look at death, class, micro-aggression, popular culture, access, free movement, stake in society, lingual fracas, masculinity, and more" (Shukla, 2016, no page). Taken together the essays provide a powerful insight into burdens of tolerance. They provide windows into what Povinelli (2006, p. 98) describes as "lifeworlds that must develop a mode of living within the brackets of recognition". By documenting the dispositions of gratitude and an eagerness to please (Okwonga, 2016, p. 232); or a constant sense of being on probation (Kam, 2016); or feeling exhausted and worn down by everyday and repeated experiences of racism and micro-aggression that come with the realisation that tolerance might not come with acceptance (Okwonga, 2016, p. 232), the essays in *The Good Immigrant* allow us to "become attuned to the tactics of world making within the waiting rooms of history that often never appear" (Povinelli, 2011, pp. 98–99). As such we can begin to comprehend and appreciate the other social worlds that are lived within, against, and amidst the 'brackets of recognition' and the kinds strategies that those living there use to maintain their social lives.

I am convinced by Brown's arguments that tolerance is a damaging and flawed discourse of power. Yet, as Brown and Forst (2014) has noted elsewhere – in a debate on the power of tolerance with Rainer Forst – tolerance remains a necessary supplement (in the Derridean sense) to equality. In liberal democracies where promises of equality are based on sameness, tolerance manages antagonisms and difference. Tolerance saves lives but it also entrenches inequality and secures the abjection of some people from the body politic. This 'paradox of tolerance' (Gentile, this volume) is where my trouble with tolerance lies. Caught between these incommensurable forces internal to tolerance, I end up getting stuck. And my way of getting unstuck involves some sleight of footwork, or perhaps a stumble, to side-step addressing tolerance directly. But while I move away from more abstract discussions of tolerance, I stay with the frictions, contradictions, and ambiguities that these different orientations to tolerance highlight, as I seek other ways of understanding and cultivating habits of being-with Others, and think about how new spaces where an ethics of conviviality can take place might be crafted.

Troubling research: research as world-making

> A revolutionary action within culture must aim to enlarge life, not merely express or explain it.
>
> (Debord, 1957, p. 36).

In the United Kingdom, talk about multiculturalism is routinely couched in narratives of failure and crisis. There's an unrelenting focus on spectacular and troubling events that include urban disorder and disturbance, terrorist attacks, gang violence, religious extremism and sexual exploitation, regressive and repressive practices that curtail the freedom and rights of women. Multiculturalism has become a variant of what Berlant (2016, p. 398) calls "the wreck of the old good life fantasy". Against this talk of crisis and failure, momentum has gathered behind research that emphasises the ordinary and unspectacular ways in which people live together in cities – much of which is manifest in an intellectual excitement about encounters. Encounters have been mobilised in more or less hopeful ways to point to the ongoing, ordinary, and unspectacular ways in which social differences are navigated and negotiated in cities. The turn to encounters in ethnographic research on living with difference in British cities has been sparked by a dissatisfaction with a perpetual handwringing in popular and political discourse about segregation, cultural difference, and the troubles associated with sharing urban space (Amin, 2002; Wilson, 2017; Swanton, 2018). Encounters engender an attunement to phenomenologies of everyday social mingling and document the acts of kindness, the comfort of rhythms, the soundscapes, the ordering, the petty tensions, the ghosts, the dreams, the humdrum violence that are part of the warp and weft of the living in multicultural cities. And research on encounters takes on the labour of documenting the everyday performances of living with difference in streets, cafés, taxis, buses, parks, busy markets, playgrounds, and neighbourhoods, and of theorising the importance of micro-publics for unsettling prejudices and nurturing moments of tolerance, engagement or recognition (Amin, 2002).

In geography, sociology, and anthropology the concept of encounter has emerged, then, as a rallying point for research that seeks to get to grips with, and make sense of, what Doreen Massey (2005) called sites of 'throwntogetherness' where social differences are made, negotiated, and contested. Cities attract and bring together people from different parts of the world with different histories, biographies, and journeys; who embody or subscribe to different social and cultural identities; who express multiple forms of attachment and belonging; who occupy different class positions (Leitner, 2012). Much of the research seeking to understanding the throwntogetherness of urban life comes under the umbrella of geographies of encounter – "a shorthand for a body of work broadly interested in social diversity, urban difference, and prejudice, which has sought to document how people negotiate difference in their everyday lives" (Wilson, 2017, p. 45). A leitmotif of research on urban encounters has been ethnographic attention to the micropolitics of everyday life in particular multicultural places (Amin, 2002; Wise and Velayutham, 2009). This attention to ethnographic detail and the micropolitics of daily life has been instrumental in shaping claims about the significance of understanding prosaic spaces of urban encounter, and the identification of the ways in which spaces like evening classes, publics transport and urban gardens might offer up opportunities for negotiating across ethnic boundaries to nurture forms of agonistic multicultural conviviality and craft spaces of tolerance (Amin, 2002). This work on geographies of encounter has inspired a growing body of research that is concerned with understanding the phenomenology of everyday urban life and how social differences are experienced, negotiated, and contested. Fashioned against more abstract claims about forms of cosmopolitanism fostered by interaction in public space, research on urban encounters research points to the messy, unruly, and at times contradictory relations of that make up urban multiculture in the shared spaces of the city.

Quite rightly there has been some reflection on the claims made for the possibilities of encounters between strangers. Gill Valentine (2008) has warned against overly celebratory claims that encounters might form a foundation for a politics of living with difference. She notes that good or positive encounters often fail to travel, and that encounters often leave attitudes unmoved, or worse still can entrench prejudices. However, the emphasis on encounters offers routes to an agonistic politics of living with diversity based on engagement and negotiation, and a welcome alternative to the existing policies that stress social cohesion based on fixed notions of common values, community and shared identifications (Fincher and Iveson, 2008). My sense is that these appeals to encounter and agonistic politics are a little too comfortable. Helga Leitner (2012, p. 842) suggests that "spaces of encounter show us that categories of difference can be destabilised so that new spaces for negotiating across difference can be created". However, the attempts of geographers and others to create these kinds of spaces often seem vague and timid, largely limited to debates about the design of urban public spaces. My frustration is that important scholarly writing on encounters misses opportunities to make more of an impression in the world. One of the reasons for this is that social scientists have tended to theorise encounters in narrow ways that focus on documentation and description for the purpose of interpretation and analysis. But if we embrace the idea of performativity, and a conviction that representations do things (Anderson, 2018), then we can begin to think about how the work of documenting and describing the lived experiences of urban multiculture opens up ways of being more experimental and political.

In the examples discussed in this chapter, I develop the notion of infrastructures for living with difference to develop a 'spatial grammar' that is concerned with taking the place of politics, and I look to develop how encounters and situations might be understood as political. Drawing on some recent writing on infrastructures, I want to sketch the contours of 'infrastructures for living with difference' that uses displacement and disruption to unlearn habits of race thinking, reboot forms of sociality, and reimagine forms of affective continuity that involve new attunements to being with Others. Along the way, I hope to respond to – aspire to – Kim Fortun's recent argument for experimental ethnography:

> Our task now becomes creative. We must try, through the design of an experimental ethnographic system, to provoke new idioms, new ways of thinking, which grasp and attend to current realities. Not knowing in advance what these idioms will look and sound like.
>
> Ethnography must, then, create a space for deliberation, for worrying through, for creativity. It must stage encounters. Operating a bit like performance art.
>
> (Fortun, 2012, p. 453)

Infrastructures for living with difference

Over recent years geographers, anthropologists, and others have used infrastructures (and their failure) as a lens to understand the organisation and experience of contemporary life (Easterling, 2014; Graham and Marvin, 2001; Graham, 2010; Anand, Gupta and Appel, 2018). Much of this work has focussed on the hidden material networks of pipes and cables that form the background of life to understand the material processes and relations through which uneven and unequal worlds are made and sustained. Infrastructures are sites where politics is translated from rationality to practice through the provisioning (or not) of the means for social and physical reproduction, and as the terrain where promises, ethics and political authority is negotiated (Easterling, 2014). The emphasis is often on accounting for the "participation of non-human forces in the organisation, practice and political life" (McCormack, 2017b, p. 420), and on recognising that infrastructures are achievements forged through practices of engineering, building, and improvising.

The geographer Derek McCormack (2017b, p. 418) has recently argued that we need to understand infrastructures not only as technical arrangements but also as "organised assemblages of material, technology and practice generative, in turn, of different worlds, or of what Kathleen Stewart (2014) calls 'registers' of intimacy and alienation". In this vein, Lauren Berlant (2016, p. 393) has argued that "infrastructure is defined by the movement or patterning of social form. It is the living mediation of what organises life: the lifeworld or structure". These interventions point to some exciting new directions for thinking about infrastructure, opening horizons for thinking about infrastructure beyond the provisioning of utilities and services. For example, Berlant (2016) reads the commons as an infrastructure for troubling times. She uses infrastructure to make sense of but also imagining ways of living in, what she calls the 'glitch': a present of infrastructural breakdown as modern practices and promises of resource redistribution, social relations, and affective continuity fall apart. Here, Berlant points to the importance of building affective infrastructures and imagining new patterns, habits, and norms of sociality.

Borrowing and building on this recent writing on infrastructures I develop a spatial grammar that is concerned with imagining how we might understand, experiment with, and modify infrastructures for living with difference. In particular, I focus on more-than-representational forms of politics that emphasise the doing of being with others and the taking place of conviviality. In my first example, I am interested in focussing on Simone's arguments for understanding 'people as infrastructure'. Through a makeshift memorial to victims of Far-Right Terror in Christchurch that is improvised amongst the railings of Edinburgh's Central Mosque, I am interested in the kinds of repair work that takes form in the midst of racisms ruins. Here, I use infrastructure and 'broken world thinking' (Jackson, 2014) to point to different modes of appreciating and understanding ongoing social projects and worlds that involve building and maintaining connections and solidarity as a situated ethics of conviviality. Inspired by Anna Lowenhaupt Tsing's (2016, p. 2) arguments for the possibilities of life in ruins, I want to follow her and take up the "imaginative challenge of living without those handrails, which once made us think we knew, collectively, where we were going". While Tsing's project involves opening herself to the fungal attraction of the matsutake mushroom as she seeks to make sense of precarious times, I borrow her method of listening for, and telling, rushes of stories that embrace the patchiness of entangled lives and points to the importance of modern of writing that craft "opportune occasions for taking up and being taken up by the 'force that opens up another world, a force that is immanent to the existing world" (Cheah, 2008) (McCormack, 2017a, p. 10). In my second example that examines the performance encounters sage through Mahtab Hussain's photographic portraits of British Muslim men, I am interested in the more purposeful engineering and building of infrastructures for living with difference. My argument is that Hussain's project You Get Me? produces spaces of encounter with young British Muslim men in ways that call out racist visual cultures fed by negative media coverage. These portraits stage urban encounters that have the potential, I argue, to disrupt practices of looking and sorting that are entrenched in habits of race thinking; and they offer the ground for imagining new patterns, habits, and norms of sociality.

Repair work: a makeshift memorial

Outside the Edinburgh Central Mosque, a scene gathers. A makeshift memorial takes form in the hours and days after the murder of 50 worshippers in two mosques located on the other side of the world in Christchurch. An assemblage of flowers, cards, cartoons circulating on Instagram, prayers, candles, Buddhist prayer flags, and hashtags sprout and proliferate amongst the ornate railings that separate the mosque from a busy street. This assemblage of material things in this place – and in many other cities around the world – points to what Berlant (2011, 15) would call "scenes of affective adjustment to material that meditates the ongoing present across the recent, the now and the next". Becoming attuned to this scene

> provides a technique of glimpsing how worlds show up as the emergent outcome of arrangements of things, devices, props, etc. A stage or a set is a

machine for producing the aesthetic scenographic affects of a world, and, what is more, those who witness the production of these affects often know it: they are not dupes.

(McCormack, 2017a, 9)

Unhappily such scenes have become familiar as public expressions of the emotional responses and affective adjustments to terror attacks across the world. But taking my cue from writers like Kathleen Stewart (2007) and Lauren Berlant (2011) I examine how we might become attuned to the affective atmospheres, and the something that is happening, in the scene of this memorial and my encounter with it. Stewart and Berlant exemplify a mode of writing that alerts us to the politics of the more-than-representational. In *Cruel Optimism* Berlant powerfully shows how affective atmospheres – and our habits of adapting to them are political. She writes: "the visceral response is a trained thing, not just an autonomic activity. Intuition is where affect meets history in all of its chaos, normative ideology and embodied practices of discipline and invention" (Berlant, 2011, p. 52).

In the midst of racism's ruins, something hopeful surfaces.[1] This makeshift memorial responding to the violence and horror of the attack in Christchurch opens up possibilities for living better with difference, surfacing through expressions and messages of sympathy and solidarity. This memorial registers a shift in the mood

Figure 4.1 Makeshift memorial, Edinburgh Central Mosque

and stance of towards minority social projects and worlds. In this assemblage of objects we find the material traces of Back and Sinha's (2016) 'situated ethics of conviviality'. Rather than scripting these scenes as spaces of tolerance, it might be instructive to reflecting on how these spaces enable the cultivation of an ethics of care for the how of every situation (Braun and Wakefield, 2014, p. 10). In the wake of the mosque attacks this place quickly emerges as a site of repair work. As soon as worlds fall apart – and the handrails (multiculturalism, tolerance, cosmopolitanism) that once offered reassurance disappear – so the work of holding together and constructing an alternative habitus starts to take place (Vasudevan, 2015).

LOVE FOR ALL
HATRED FOR NONE
#UNITEDAGAINSTRACISM

We love you and will keep you in our prayers for ever. May your souls rest easy ♥ ♥ ♥ ♥
We're sorry for what happened in NZ and we stand with you our Muslim brothers and sisters.
Love & prayers
♥
Kiwi girl in Edinburgh

Figure 4.2 "This is your home and you should have been safe here"

The makeshift memorial is an example of a deathscape. The mosque in Edinburgh becomes one in a series of

> places that have or take on meaning in relation to the dead [and] can therefore act as a catalyst, evoking grief, memories, sadness and comfort – or an unpredictable combination thereof. Specific locations associated with the deceased in life or death, can also [. . .] be sites for action, the active emotional-affective practices and performances of expression, remembrance and ongoing relation, for example a roadside shrine.
>
> (Maddrell, 2016, p. 170)

While the mosque attacks took place 11,500 miles away from this street in Edinburgh the site takes on a sense of meaning. In this memorial, we witness an immediate outpouring of grief. And this grief is shared in ways that produce and reinforce a sense of a collective. This collective is manufactured through connections made through the performative force and repetition of hashtags like #unitedagainstracism, or through the performativity of the expression 'We stand with you our Muslim brothers and sisters'. Here there is a sense of people rising up to a disaster, and feeling connected throughout the horror of the mosque attacks. Feelings of empathy, sorrow, love, and solidarity come to the fore. But so too does a sense that something has to be done. The violence of the attacks something becomes sensible and has to be negotiated as an ethical substance. These murders unmask an unequal distribution of life and death, harm and hope across social difference. It lays bare the vulnerability of migrant communities and religious minorities. For many, the horror of the attacks in Christchurch demand a response that produce concrete acts and contexts that suddenly appreciate social worlds lived and constructed in the 'brackets of recognition' (Povinelli, 2011) and to build new (often fleeting) connections and forms of solidarity.

Rebecca Solnit (2005) identifies something similar after September 11, 2001 in a short essay in *Hope in the Dark*. She acknowledges the horror of the terror attacks, but also reflects that "something beautiful might have come out of it". Alongside belligerent and racist reactions, "there was a long moment when almost everyone seemed to pause, an opening when the nation might have taken another path" (Solnit, 2005, p. 54). Solnit is interested in the ways in which people rise to the occasion in a disaster, and how this offers hope. She points to the impulse to donate blood, to depart from the habitual rhythms and routines of everyday life by staying home from work, and by talking to strangers. She also hopes that the terror attacks dislodge a sense of American invulnerability and foster a stronger commitment to global citizenship. Solnit sees such acts as modest forms of heroism. Engendering 'selfless states of being' and a 'willingness to do' Solnit argues that disasters open up the potential for people to make connections and commit to community, transgressing the habits of relating to Others through dispositions that range from preoccupied indifference to suspicion to manifest hostility. In the process, we witness how personal feelings of sympathy and solidarity can spark forms of collective action that might be more readily recognised as 'civic solidarity' in liberal theories of democracy.

This roadside memorial offers an opportunity to work with the everyday experiences, encounters and actions that surface in the wake of a terror attack. Rather than retreating to abstract discussions of how tolerance and intolerance might explain the actions of a Far-Right terrorist, the contexts that gave rise to his crimes, or how we should collectively respond to them, this memorial opens a politics that 'returns to the world'. Engaging with 'ruins-thinking' attunes us to how people live in situations that are uncertain, even volatile. Focussing on the memorial helps me muddle through the attacks in Christchurch and the wakes they have produced, sticking with the hard, the messy and the messed up (Berlant, 2016). Amidst racisms ruins and collapsing popular support for multiculturalism, we might learn from others who live with ruins, amidst ruined fantasies of a good life. The memorial at Edinburgh's mosque points to the ways in which people improvise and make do (Simone, 2004; Vasudevan, 2015). They make use of the infrastructures, objects, ideologies of memes that are around them. People borrow from what is around them and experiment on their own terms to make connections, forge relations, and build solidarities, however fleeting. The memorial embodies a 'here and now' mentality (Wakefield, 2018) that is a collective exercise in world-making. The possibilities of convivial and collective life do not reside in dreams of some 'enchanted world' but by confronting the destructions of such dreams in the violence of a terror attack (Berlant, 2011; Wakefield, 2018). The memorial registers shifting affective attunements to Others. It points to new ways of being with Others, and the work of weaving new connections and relations with other and the world. In the ruins of racisms, the memorial begins to remake the city, emphasising collective capabilities of making multiculture, where conviviality and solidarity coexists with – and even takes shape through – racism and its ruins.

The example of this makeshift memorial diverts attention from abstract question of tolerance that have circulated around the mosque attacks. It focusses rather on cultivating an ethics of care for the how of every situation (Braun and Wakefield, 2014, p. 10). It echoes Lauren Berlant's (2011) project of building something "akin to a how-to guide for living in the impasse". What will we use, what will we leave idle, what will we destroy? How will we take care of each other? How will we call out the humdrum violence of racism and Islamophobia in public life? How will we shift senses of entitlement? How will we short-circuit discourses that dramatise differences and denigrate Others? How might we exploit glitches in race thinking? How will we dwell in the ruins of racism?

In my second example, I shift from a reading of a gathering scene to a focus on a more purposeful and engineered example of building infrastructures for living with difference. In her argument for the importance of understanding 'affective infrastructures of the ordinary', Berlant tunes us into situations in ways that help us see how the affectivity of a moment is always about the construction of social norms, conventions and rules, fantasies of the good life. But more importantly,

Berlant insists that our responses to affective atmospheres are learnt. As such, new spaces of political action become possible. If visceral responses are trained, they can also be disrupted, subverted, or retrained. And so we might look to construct moments of impasse – suspending animation – to experiment with alternate ways of relating, being with, making sense, adapting and moving on in multicultural places. Mahtab Hussain's project You Get Me? offers one such example on constructing an impasse.

World-making: 'You Get Me?'

Performative encounters produced through arts practice provide pathways to forms of creative and communicative politics that craft new ways of being with, and relating to, Others in multicultural cities. Practices like photography, street theatre, or dérive offer ways of participating in the world that perform the political activism and collective work of what Ahmed (2000, p. 17) calls strange encounters by working "to find ways of re-encountering these encounters so that they no longer hold other others in place". The participatory, and sometimes playful, nature of performative encounters has the potential to subvert and destabilise existing uncongenial relations and entrenched habits of race thinking. Performative encounters also offer tactics for building alternative relations, dispositions, and spaces that might become part of an infrastructure of living with difference. The scene of an encounter has the potential to make folds in social relations; the 'swerve' of an encounter can make other things happen, it becomes the means for a new attunement, a new history (Berlant, 2016).

I examine how the work of photographer Mahtab Hussain, and in particular his project You Get Me?, produces performative encounters. Hussain's photographic portraits craft encounters with young British Muslim men in ways that call out racist visual cultures fed by negative media coverage. These portraits stage urban encounters that have the potential, I argue, to disrupt practices of looking and sorting that are entrenched in habits of race thinking; and they offer the ground for imagining new patterns, habits, and norms of sociality.

Photographer Mahtab Hussain's portraits capture performances of masculinity and identity by British Muslim men living in inner-city, working-class neighbourhoods. The sitters for Hussain's portraits were all approached in chance encounters. The photographs capture the men and boys Hussain encountered and talked to on the streets of British cities. The background for the portraits is the street with backdrops including shutters, shop windows, parked cars, gym equipment, low walls, and swings. This staging of the portraits is purposeful. It both situates and amplifies the performances of masculinity and identity and how these performances are shot through with concerns about power, money, respect, and territoriality. In his photographs Hussain challenged his sitters to visually articulate what they want to say to a wider audience (Gander, 2017). The force of the portraits arrives in the ways in which they catalogue diverse performances of masculinity and identity by British Muslim men. Alongside the photographs, unattributed quotes from sitters offer viewers windows into the lives lived by young British Muslim men. Their stories, struggles, and aspirations inject vulnerability and humanity into representational repertoires that repeatedly construct these men as some monstrous Other in post-colonial Britain.

You Get Me? looks to hold media and political discourse to account. Hussain's work disrupts the dominant visual regimes through which British Muslim masculinity is produced, circulated and consumed. For example, a book produced as part of the project features a striking portrait of a young man whose piercing eyes hold the viewer's gaze (Figure 4.3) while the back cover offers a catalogue of headlines from the British tabloid press that index some of the many ways in which British Muslims are Othered, vilified and cast out. The juxtaposition of front and back cover is forceful. The photograph on the front cover demands a response. I find myself returning to this image, caught in this man's gaze. There's something familiar about the portrait. A young Asian man. Hair slicked forwards with gel. A patchy beard. An earring in his left ear. A kaffiyeh prayer scarf – part of the iconography of Palestinian nationalism – draped over his shoulder. The force of media representations means that this stranger is recognised (Ahmed, 2000). Recognised as Other. But even as the photograph is situated within the motifs and visual regimes that construct British Muslim masculinity as Other, deviant, and threatening, it also punctures this visual regime. The portrait returns a look. There's something about the directness of this man's gaze that demands a response. It is as if there is a hard-wired response when we encounter the eyes of another human

Figure 4.3 "Shemagh, beard and bling" (Mahtab Hussain, *You Get Me?*, 2017)

(or non-human) being. Held in this sitter's gaze, we cannot not want to know his story. Who is he? What are his hopes? His loves? His dreams? His fears? This work of puncturing and disrupting practices of looking and sorting are an important part of what Hussain's photographs want. In an interview with the *Independent* newspaper he frames the project as one that challenges what viewers are looking at:

> Do you see men? Are they British men, Asian men, Muslim men? Can you dare define them as fine art portraits, (because that is where I stand), worthy to hold such a little, or are they documentary, because black and brown men photographed in their own space by [a] black and brown artist can be nothing but documentary.
>
> (*The Independent* 23.05.2017)

One reading of You Get Me? is as a project of deconstruction. It intervenes in the image-repertoire through which British Muslim men are routinely represented. Claire Alexander (2000, 2004) has written powerfully about how moral panics about Asian gangs, and more recently 'home-grown terrorism', have made British Muslims hypervisible in the media. News media produce a near daily bombardment of negative coverage that constructs British Muslim men as terrorists, gang-members, ISIS sympathisers, and sexual predators. This coverage is accompanied by photographs of brown bodies: grainy stills from CCTV footage; police mugshots; photographs gleaned from social media or snapped in the street. These circulating photographs resemble bodies that we encounter in the course of our everyday lives. Images format perceptions and shape dispositions in ways that come to mediate encounters (Swanton, 2010).

The portraits in You Get Me? carry a deconstructive force. The photographs and quotations intervene and disrupt; they return looks and speak back to challenge the visual regimes through which young working-class men in Muslim communities are represented in the media. But these portraits are also objects in the world. They take the form of framed prints exhibited in gallery spaces; or as images printed in newspaper supplements or in coffee table books; or as pixels and digital files posted and shared via websites and social media. As objects in the world these portraits that travel and circulate have a performative force. While images always appear in some kind of material form (film, paint, toner, pixels) a crucial feature of the lives of images is their ability to circulate from one medium to another, to move from the page to the screen, from the screen to the performances of everyday life. Introducing his examination of what pictures want, Mitchell (2005, p. xv.) argues that pictures are " 'ways of worldmaking,' not just world mirroring". Mitchell is interested in a poetics of pictures that explores 'the lives of images', sidestepping questions of what pictures mean or do to dwell on "what they want – what claim they make upon us and how are we to respond?"(ibid). Mitchell shifts our orientation to pictures, and by focussing on the social lives of photographs allows us to consider the ways in which Mahtab Hussain's portraits have the potential to produce a swerve in everyday urban encounters. The photographs become "go-between's in the social field of the visual, and everyday processes of looking and being located" (Mitchell, 2005, p. 47).

Producing and curating portraits of British Muslim men, Hussain crafts uncertain, creative, and sometimes playful urban encounters that disrupt, displace, and engineer patterns, habits, and norms of sociality. Playing on motifs of swagger, surprise, machoism, vulnerability, and struggle, the images call out racist stereotypes, and provide a medium for subverting uncongenial relations and entrenched habits of race thinking in British cities. I argue that these portraits stage performative encounters that explode one-dimensional media representations that transact in stereotypes by exposing the viewer to the life and liveliness of young British Muslim men. In the process, the portraits provide new ways of relating to others, and introduce a glitch into the comfortable habits and social norms of middle class multiculture. While staging encounters, You Get Me? also troubles habits of looking, sorting, and judging difference in cities, inviting what Butler and Anthanasiou (2013) call 'responsiveness as responsibility'.

Staging encounters

Hussain's portraits are staged in the settings of his sitters' everyday lives. The backgrounds of the portraits are the ordinary streetscapes of working-class, inner-city neighbourhoods. Many of the portraits are framed by the cluttered materiality of urban streetscapes (signs, shop fronts, shutters, garage doors, low walls, parked cars, weeds, and groomed hedges), others take place against the backdrop of backyards, scruffy urban parks, gyms, boxing clubs, and cars. This framing of the portraits was purposeful. It locates these men, and conveys something of the spaces in which their performances of masculinity and identity unfold. In a very real sense, this framing stages urban encounters, and looking at the photographs will transport some viewers to unfamiliar neighbourhoods. These are neighbourhoods that some would not ordinarily venture into as their navigation of the city is contorted by popular imaginative geographies that construct them as scary, as no-go areas, as 'Muslim ghettos'.

Beyond locating his sitters through the framing of his portraits, Hussain performs the important work of exposing the deadening effects of the visual cultures through which British Muslim men are routinely encountered. Mitchell (2005, p. 298) argues that racial stereotypes are defined by "precisely what they lack – life, animation, vitality". You Get Me? captures some of the life and liveliness of British Muslim men's identifications, attachments, and performances that is snuffed out by the mugshots or grainy stills that dominant media representations. While these men are clearly acting up to the camera – posing and dramatising particular aspects of their daily performances – the portraits evoke the multiple, contradictory, and messy lives of British Muslim men.

The portraits do not shy away from youth cultures that are entangled with moral panics about gangs. In a number of the portraits, boys and young men are clearly playing up to typecasts of the big man; their performance for Hussain's camera involves projecting 'gangster swagger' expressed through a gestural economy of gait, pose, and hand signals that projects a confidence shaped by their social location. In addition to performances that play on respect and power, there are other portraits that capture these men at work, hanging out, eating, walking dogs, working out, etc. The portraits show the many ways in which social distinctions are performed through style. Style is manifest in the careful choice of clothing that hints at diverse attachments and identifications: many wear designer sportswear, others pose bare-chested or in vests chosen to show off their gym-honed torsos, one boy wears a T-shirt emblazoned with an anti-racist slogan, another is dandy-esque in a tailored three-piece suit, and others pair salwar kameez with hoodies and Adidas sneakers. The portraits also introduce material cultures of car enthusiasm, cultures of body-image that aspire to gym-sculpted bodies (and perhaps the consumption of performance enhancing drugs), and the status and respect that is earned through the ownership of aggressive breeds of dogs. The style, comportment, and material cultures of British Muslim masculinity on display in these portraits are diverse and full of life. The portraits capture aspects of lives lived and stories to be told through the hybrid material cultures of these neighbourhoods. They tell stories of how status, power, and respect are achieved in working-class urban neighbourhoods.

In the process the portraits stage an impasse. They introduce a glitch into the comfortable lives of middle class urban multiculture. These portraits evoke scenes of 'cruel optimism', where "we are forced to suspend ordinary notions of repair and flourish to ask whether the survival scenarios we attach to those affects weren't the problem in the first place"(Berlant, 2011, p. 49). So while it is important to trouble some of the performances of masculinity and identity in the images, they speak more profoundly to forms of the vulnerability, and the daily struggles these men experience. They bring into focus relations of living on in relation to crisis, loss, and hardship. Hussain forces home how these visual performances often mask vulnerabilities, wounds and struggle by including quotes from the men who sat for him. Many of which dwell on their representation in then media after 9/11. These portraits suspend normal ways of relating and making sense. They attune us differently to situations, perhaps retraining affective responses.

Conclusion

In this chapter I have sought to trouble the place of tolerance in thinking about urban multiculture in British cities. Public and political discourse is often quick to retreat to abstract and shifty values of tolerance in response to the messy and messed up events and lived realities of urban multiculture. Political theorists have been particularly influential in shaping how the tolerance of religious and cultural difference is negotiated and established in contemporary liberal democracies (Jones, 2007; Gentile, this volume). However, my argument is that when approached from spaces of everyday multiculture and a situated ethics of conviviality, these ideas and ideals of tolerance feel abstract and limited. They require that the difficulties, messiness, and complexity of day-to-day lives be suspended. My interest in this chapter has been to open a different orientation to the scenes, moments, and events that often spark tolerance talk. Foregrounding the ethnographic details of everyday spaces of tolerance (alongside other affective orientations to difference), and the ongoing negotiations of difference, has allowed me to reflect on how new attunements to difference and ways of relating to Others might be nurtured out of difficult and messed up moments of everyday multiculture.

Through my examples of a makeshift memorial in the wake of terror attacks in Christchurch in March 2019 and photographer Mahtab Hussain's project You Get Me?, I have sought to exemplify how recent theorisations of infrastructure can help develop a new spatial grammar for understating a situated ethics of conviviality (Back and Sinha, 2016) in ways that think anew about spaces of tolerance. The example of the makeshift memorial points to the repair work that took place in the immediate aftermath of a Far-Right act of terror. The memorial offers material traces of tentative practices of making connections and imagining new modes of being-in-common that emerge from the ruins of racisms and destruction of cosmopolitan dreams. Possibilities of collective and convivial life are seeded by confronting the violence of these terror attacks. You Get Me? provides an example of how encounters produced through works of art might manufacture spaces in which differences might be encountered in new ways. Social scientists have a lot to learn from the many methods that artists use to intervene in urban life, and to unsettle experience through alternative stories and histories. Collaborating with, and learning from, arts practices provides spaces, methods, and tactics for producing performative encounters that might unsettle social norms and entrenched habits of (racialised) sense-making. Performative encounters have the potential to spark public imagination, catalyse political contestation, and produce engaged publics. There is a politics and poetics in Hussain's photographs that acknowledges what is unseen, overlooked, or neglected, while at the same time working ways of knowing and representing that seek to rewrite geographies and make urban worlds anew. And if we are interested in building infrastructures for living better with difference, we might look to other possible lives that might be constructed in the ruins of racism. How might poems, books, photography, plays, films, lessons, workshops, sports games, walking tours, social sculpture, amongst other things intervene in multicultural life? How might they make available resources, ways of thinking, material objects, processes, spaces that unsettle disrupt, haunt, and trouble? And how might they then allow us to reflect on our experiences and the experiences of others; to make connections; to take advantage of opportunities to collaborate; and to build, repair, and maintain solidarity and collective life as part of a situated ethics of conviviality?

Note

1 See Bialasiewicz (2019) for a discussion of the ordinary discourses around existential threats to white-Western-European civilisations that underpin exceptionally violent events like the Christchurch mosque attacks, but also incubate and normalise 'far- (and not-so-far) right' political movements in the United States and Europe.

References

Ahmed, S. (2000). *Strange encounters: Embodied others in post-coloniality*. London: Routledge.

Alexander, C. (2000). *The Asian gang: Ethnicity, identity, masculinity*. London: Berg.

Alexander, C. (2004). Imagining the Asian gang: Ethnicity, masculinity and youth after 'the riots. *Critical Social Policy*, 24, pp. 526–549.

Amin, A. (2002). Ethnicity and the multicultural city: Living with diversity. *Environment and Planning A*, 34, pp. 959–980.

Anand, N., Gupta, A. and Appel, H. (2018). *The promise of infrastructure*. Durham, NC: Duke University Press.

Anderson, B. (2018). Cultural geography II: The force of representations. *Progress in Human Geography*, pp. 1–13.

Back, L. and Sinha, S. (2016). Multicultural Conviviality in the Midst of Racism's Ruins. *Journal of Intercultural Studies*. 37(5), pp. 517–532.

Barnett, C. (2008). Political affects in public space: Normative blind-spots in non-representational ontologies. *Transactions of the Institute of British Geographers*, 33, pp. 186–200.

Barnett, C. (2012). Situating the geographies of injustice in democratic theory. *Geoforum*, 43, pp. 677–686.

Berlant, L. (2011). *Cruel optimism*. Durham, NC: Duke University Press.

Berlant, L. (2016). The commons: Infrastructures for troubling times. *Environment and Planning D: Society and Space*, 34, pp. 393–419.

Bialasiewicz, L. (2019). Ordinary ideas: Demographic geopolitics and the imagined battle for Europe. *OpenDemocracy*. Available at: https://www.opendemocracy.net/en/can-europe-make-it/ordinary-ideas-demographic-geopolitics-and-imagined-battle-europe/

Braun, B. and Wakefield, S. (2014). Governing the resilient city. *Environment and Planning D: Society and Space*, 32, pp. 4–11.

Brown, W. (2006). *Regulating aversion: Tolerance in an age of identity and empire*. Princeton, NJ: Princeton University Press.

Brown, W. and Forst, R. (2014). *The power of tolerance*. New York: Colombia University Press.

Butler, J. and Anthanasiou, A. (2013). *Dispossession: The performative in the political*. Cambridge: Polity Press.

Cheah, P. (2008). What is a world? On world literature as world-making activity. *Daedalus*, 137(3), pp. 26–38.

Debord, G. (1957). *Report on the construction of situations*. Available at: https://theanarchistlibrary.org/library/guy-debord-report-on-the-construction-of-situations.pdf

Easterling, K. (2014). *Extrastatecraft: The power of infrastructure space*. London: Verso.

Fincher, R. and Iveson, K. (2008). *Planning and diversity in the city: Redistribution, recognition and encounter*. Basingstoke: Palgrave Macmillan.

Fortun, K. (2012). Ethnography in late industrialism. *Cultural Anthropology*, 27(3), pp. 446–464.

Gander, K. (2017). You get me? Photographer captures young British Asian Muslim men to break down stereotypes. *The Independent*. Available at: http://www.independent.co.uk/life-style/you-get-me-mahtab-hussain-british-asian-muslim-men-photos-stereotypes-negative-a7751056.html. [Accessed 13 Feb. 2018].

Gill, N., Johnstone, P. and Williams, A. (2012). Towards a geography of tolerance: Post-politics and political forms of toleration. *Political Geography*, 31(8), pp. 509–518.

Graham, S. (2010). *Disrupted cities: When infrastructure fails*. London: Routledge.

Graham, S. and Marvin, S. (2001). *Splintering urbanism: Networked infrastructures, technological mobilities and the urban condition*. London: Routledge.

Hage, G. (1998). *White nation: Fantasies of white supremacy in a multicultural society*. Wickham: Pluto Press.

Hussain, M. (2017). *You get me?* London: Mack.

Jackson, S. (2014). Rethinking repair. In: T. Gillespie, P. Boczkowski, and K. Foot, eds., *Media technologies: Essays on communication, materiality and society*. Cambridge: MIT Press, pp. 221–239.

Jones, P. (2007). Making sense of political toleration. *British Journal of Political Science*, 37, pp. 383–402.

Jones, P. (2012). Toleration, religion and accommodation. *European Journal of Philosophy*, 21(3), pp. 542–563.

Jones, P. (2018). *Essays on toleration*. London: ECPR Press/Rowman & Littlefield.

Kam, W.M. (2016). Beyond 'good' immigrants. In: N. Shukla, ed., *The good immigrant*. London: Unbound.

Kim, C.J. (2015). *Dangerous crossings: Race, species and nature in a multicultural age*. Cambridge: Cambridge University Press.

Leitner, H. (2012). Space of encounter: Immigration, race, class and the politics of belonging in small-town America. *Annals of the Association of American Geographers*, 102(4), pp. 828–846.

Lentin, A. and Titley, G. (2011). *The crises of multiculturalism: Racism in a neoliberal age*. London: Zen Books.

Maddrell, A. (2016). Mapping grief: A conceptual framework for understanding the spatial dimensions of bereavement, mourning and remembrance. *Social and Cultural Geography*, 17(2), pp. 166–188.

Massey, D. (2005). *For space*. London: Sage.

McCormack, D. (2017a). The circumstances of post-phenomenological life worlds. *Transactions of the Institute of British Geographers*, 42(1), pp. 2–13.

McCormack, D. (2017b). Elemental infrastructures for atmospheric media: On stratospheric variations, value and the commons. *Environment and Planning D: Society and Space*, 35, pp. 418–437.

Mitchell, W.T.J. (2005). *What pictures want: The lives and loves of images*. Chicago: University of Chicago Press.

Modood, T. and Dobbernack, J. (2011). A left communitarianism? What about multiculturalism? *Soundings*, 48, pp. 54–64.

Okwonga, M. (2016). The ungrateful country. In: N. Shukla, ed., *The good immigrant*. London: Unbound.

Povinelli, E. (2006). *The empire of love: Towards a theory of intimacy, genealogy and carnality*. Durham, NC: Duke University Press.

Povinelli, E. (2011). *Economies of abandonment: Social belonging and endurance in late liberalism.* Durham, NC: Duke University Press.

Shukla, N. (2016). *The good immigrant.* London: Unbound.

Simone, A. (2004). People as infrastructure: Intersecting fragments in Johannesburg. *Public Culture,* 16(3), pp. 407–429.

Solnit, R. (2005). *Hope in the dark: Untold histories, wild possibilities.* Edinburgh: Canongate.

Stewart, K. (2007). *Ordinary affects.* Durham, NC: Duke University Press.

Stewart, K. (2014). Road registers. *Cultural Geographies,* 21(4), pp. 549–563.

Swanton, D. (2010). Sorting bodies: Race, affect and everyday multiculture in a Mill Town in Northern England. *Environment and Planning A,* 42, pp. 2332–2350.

Swanton, D. (2018). Urban encounters: Performance and making urban worlds. *Journal of Urban Cultural Studies,* 5(2), pp. 229–248.

Tsing, A.L. (2016). *The mushroom at the end of the world: On the possibility of life in capitalist ruins.* Princeton, NJ: Princeton University Press.

Valentine, G. (2008). Living with difference: Reflections on geographies of encounters. *Progress in Human Geography,* 32(3), pp. 323–337.

Vasudevan, A. (2015). The makeshift city: Towards a global geography of squatting. *Progress in Human Geography,* 39(3), pp. 335–356.

Wakefield, S. (2018). Infrastructures of liberal life: From modernity and progress to resilience and ruins. *Geography Compass,* 12(2).

Wilson, H. (2017). On geography and encounter: Bodies, borders, and difference. *Progress in Human Geography,* 41(4), pp. 451–471.

Wise, A. and Velayutham, S. (2009). *Everyday multiculturalism.* London: Palgrave Macmillan.

Part II

Securing and securitising religious tolerance

5 Religious toleration and securitisation of religion

Sune Lægaard[1]

Religion is both an object of processes of securitisation and of toleration. Securitisation of religion is often linked to threats from abroad, e.g. in relation to religiously defined immigrants or religiously motivated terrorism. A vivid example of both of these processes occurring at the same time happened during the 2015 refugee crisis in Europe, where migrants coming from the Middle East were mainly discussed in Europe as Muslims, and where Europe at the same time experienced terrorist attacks by avowedly Islamist perpetrators. This resulted in a situation where religion in general, but Islam and Muslims in particular, were linked to issues of security and a challenge to the ability of Europe and European nation-states to uphold their borders.

In this chapter, I will discuss what securitisation of religion implies in relation to religious toleration. Religious toleration and freedom of religion are standard ways of handling religious differences and disagreements within liberal democracies. What does it mean for religious toleration when religion is not only a matter of difference and disagreement but also an issue of security? The chapter investigates this at a philosophical level by considering the standard concepts of religious toleration and rights to freedom of religion as discussed in liberal political theory.

I consider securitisation as a discursive process in a way inspired by securitisation theory within international relations studies. However, I extend the notion to cover a broader range of phenomena like the ones considered here. I am not here arguing for the empirical claim *that* religion is securitised – this is taken as granted for the sake of argument. Rather, I ask the philosophical question what securitisation of religion *implies* for religious toleration and freedom of religion, given that securitisation is taking place. This is partly a conceptual issue, partly a normative one. I argue that securitisation makes a difference for religious toleration in both respects. Securitisation links up with several of the defining features of religious toleration, thereby making a difference for when religion becomes a possible object of toleration, and for where the limits of religious toleration are drawn. However, securitisation can also make a difference for the understanding of the different components of toleration as such and thereby change the very meaning and function of toleration. To the extent that this happens, securitisation raises a number of normative issues – because toleration in the traditional sense has served specific functions in relation to securing political and social legitimacy, which securitisation may undermine.

Securitisation and religion

Securitisation in a broad sense denotes how political actors lift some issues out of ordinary politics into extraordinary politics. The standard example is war or similar existential threats to a society: Faced with such threats, all other political issues and disagreements can often be set aside, and considerations that ordinarily limit other uses of political power typically do not apply to measures to counter existential threats. Securitisation raises the stakes and introduces a state of exception allowing for political measures not considered permissible under ordinary circumstances.

Here I will discuss securitisation processes involving religion in the sense that religion is framed as a threat (for discussion of the opposite case of securitisation of religion as being under threat, see Laustsen and Wæver, 2000). This is a somewhat weaker and broader sense of securitisation than the one established in international relations theory (e.g. Balzacq, 2010), where the threats in question are existential and the measures called for are usually of a very radical nature (eradication and elimination of the threat). In the types of cases I will consider, things have usually not come this far (yet). I will mainly consider cases where religion is linked to immigration. This is because the main reason why religion has returned to the European political agenda has to do with immigration, mostly from countries in the Middle East, south and central Asia, and Africa with predominantly Muslim populations; although a number of Islamist terrorist attacks have played an additional part. Debates about religion in Europe are often really about Islam and Muslims – and even when these debates are generalised and involve other religions and religious groups as well, the reason why the debates have taken off usually have to do with the increasing presence of and political unease with Muslim immigrants and their descendants.

These kinds of cases are not usually (except by increasingly influential radical right movements) approached politically as existential threats calling, e.g. for elimination of all Muslims. Nevertheless, it makes sense to think of religion as involved in processes of securitisation in a broader sense than the standard one from international relations theory. Therefore, when I write of securitisation of religion in the following, it denotes processes where political actors single religion out for special political concern. This opens the way for political measures in relation to religiously defined objects that would otherwise not be considered permissible (for others who use the term in this broader way, see e.g. Kaya 2012; Lægaard, 2013 for the same point without the term). While not necessarily a shift to an all-out state of exception, this process does involve a shift in the terms of political action – so securitisation comes in degrees.

Securitisation is, more precisely, the process of transforming a subject into a matter of 'security' by presenting it as involving an extraordinary problem (threat, necessity etc.) requiring certain political actions. It is important to note that securitisation is not the same as there being an objective threat. Securitisation is a discursive process of framing: it is concerned with what political actors do and with the discursive effects of this. So to say that something is securitised is not necessary to say that this issue is in fact essential to objective survival. Rather, an issue is securitised when someone is successful in constructing it as an extraordinary problem calling for extraordinary measures.

Securitisation in this discursive and processual sense is interesting because of its political effects. Successfully securitised subjects receive disproportionate amounts of attention and resources and securitisation allows political actions that would otherwise not be allowed due to the raised stakes.

The interesting question in relation to securitisation is therefore not whether the claims being made about what is a threat and what is threatened are correct or not. The interesting questions rather concern:

1   who securitises (who is the securitising actor?)
2   on what issues (what is constructed as threats?)
3   for whom (what is the so-called referent object presented as under threat?)
4   with what results (what can political actors do once an issue is successfully securitised that they could not do before?)

In this chapter, I am concerned with religion as the issue that is securitised, i.e. presented as a threat to something. The referent object for this threat is usually the nations or the citizenries of a nation-state undergoing immigration of Muslims. But since this is a process going on in several countries, including in most of Europe, the referent object is sometimes generalised so that it is not just, say, the Danes or the Dutch who are under threat but Europe and all Europeans. The referent object can also be religiously defined, e.g. when Europe is equated with Christianity. However, for present purposes, I focus on religion as the threat.

As already noted, it is actually too rough to say that 'religion' is securitised. First, it is usually not religion in general that is presented as a threat but a particular religion or religious group – in the current European case, most often Islam and Muslims. Sometimes attempts are being made to narrow the threat even more, e.g. to radicalised Islamism or particular sub-groups of radicalised Muslims. It nevertheless makes sense to say that the securitisation process is more general. On the one hand, even when distinctions are made, e.g. between dangerous forms of Islam and Islam in general, the issue of the potential threat posed tends to crop up in relation to Islam and Muslims more generally. The mere fact that the distinction between Islam and Islamism has to be invoked repeatedly indicates that the evoked worry is attached to Islam and Muslims in general – why would it otherwise be necessary to make the distinction to begin with? On the other hand, discussions about this are often framed as being about religion in general, even when people arguably think of Islam and Muslims. It is probably no accident that issues about religion in the public sphere and the relationship between religion and politics have resurfaced in parallel with the increasing presence of Islam and Muslims in Europe. This development was not prompted by concerns about established majority churches or the role of other religious minorities, although these have subsequently been drawn into the debate.

Second, the immediate issue of securitisation is often not religion as such or even a particular religion such as Islam in general but a particular religious practice. A long-running example of this is the recurring debates about Islamic headscarves. Dating back to the first French *affaire du foulard* in 1989, the headscarf as a piece of clothing has continuously been linked with fundamentalism as a radical and undemocratic interpretation of Islam, which has in turn been linked with radicalisation and potential terrorism (Laborde, 2008). Something similar happened in relation to the French so-called burqa ban in 2010 and the more curious case of the so-called burkini, which was the object of several local attempts at banning following terrorist attacks in 2015. The fact that specific pieces of clothing were in all of these cases (albeit under various descriptions) picked out for banning due to links to more general worries about security is a good example of the kind of securitisation I have in mind here. Headscarves, burqas, and burkinis were not banned because they themselves posed any threat to anything but because of assumed links to a more encompassing threat supposedly inherent in Islam as such. In this sense, the entire religion and religious group is picked out as a potential threat.

Religious toleration

Religion is not only an issue that is being securitised; religion is also an issue related to toleration. Historically, the European conception of toleration developed in relation to religious differences (Forst, 2013). Religious toleration has in turn been fundamental for development of Western liberal views of society. The return of religion to the political agenda has accordingly gone hand in hand with a return of interest in toleration. In order to investigate what the securitisation of religion means for religious toleration, I will first sketch the standard notion of toleration as theorised within political philosophy. I will subsequently also sketch the notion of freedom of religion and note the complex relation between religious toleration and freedom of religion. Despite differences between toleration and rights, religious toleration and freedom raise related *political* issues due to their similar structures.

Toleration is standardly understood in political philosophy (e.g. Cohen, 2014; Forst, 2013) as a relation between an agent A, the tolerator, and a patient B, the tolerated party, where these conditions hold:

1   difference: there is some *difference* between A and B
2   objection: A *dislikes or disapproves* of B on this basis, in a way *disposing* A to *interfere* with B
3   power: A has (or believes him or herself to have) the *power* to interfere with B
4   acceptance: A also has a reason for *not interfering*, and therefore *refrains* from doing so
5   rejection: within certain *limits* to toleration

If these conditions do not hold, we would not normally describe a relation as one of toleration. This conceptual analysis of toleration applies to single actions and to more established patterns of action (where we count refraining from interfering for some reason as an action). When toleration becomes a pattern of action, e.g. because the differences, forms of disapproval, and reasons for non-interference are stable, we might think of it as a practice or even as a culture of toleration. As such, toleration can characterise relations in a society.

Note that the analysis does not presuppose a hierarchical relation between A and B – all toleration requires is that conditions 1 through 5 hold, but nothing in the analysis assumes or implies that A has to be in a superior position to B, or that the relation of toleration cannot be reciprocal. The analysis is descriptive (Balint, 2017) in the sense that it says nothing about whether and when toleration is a good or a bad thing. Forst (2013) notes this by way of saying that toleration is 'a normatively dependent concept'.

The analysis puts no restrictions on which kinds of actors can be agents of toleration. It is immediately clear that individual persons in many cases stand in relations of toleration to each other. It also seems obvious that the state, the government, or officials acting on behalf of the state can be agents of toleration. It might be argued that the state or government in a liberal society should not object to the beliefs or practices of its citizens, e.g. since one might think that a liberal state should be neutral (see Jones, 2018 for a discussion). But even if true, this is a claim about how things should be – so, as such, it does not rule out the possibility that actual states might not in fact live up to this ideal, and that they or their agents might therefore be possible agents of toleration. Furthermore, even in a neutral state, whether one understands neutrality in terms of justification, intent, or treatment (Balint, 2017; Patten, 2014), there are bound to be things that the state can and should disapprove of because they are wrong for neutral reasons – e.g. criminal acts or inegalitarian social norms – and accordingly state toleration still has a potential role to play within liberal ideal theory (Lægaard, 2015a).

The analysis of toleration is general. Religion can enter into several of the components, so there are several different ways in which one might talk about 'religious toleration' (just as there are different ways in which religion might be involved in processes of securitisation). For present purposes, I am concerned with religion as an *object* of toleration. So religion is relevant for components 1 and 2, i.e. as some salient form of difference, which gives rise to objections. This captures the kinds of cases I am concerned with here, e.g. the presence of Islam in the public sphere in European societies, Muslim immigrants coming to Europe, or religious practices such as the headscarf or burqa associated with Islam. Religion might alternatively, or also, be a reason for or against toleration, i.e. be relevant for components 4 and 5. An example of this might be Christian charity as a reason for toleration – but I will not enter into discussion of such cases here.

On this basis, it should already now be clear that securitisation can link up with toleration. If religion is securitised, i.e. presented as a threat, this can generate objections to religion. Either the perception of religion as a threat is a distinct reason for objection to religion, or it can boost an already existing dislike or disapproval of religion. Furthermore, securitisation can also be a factor in relation to the rejection component. If religion is presented as a threat, this might be a reason why it is not tolerated, despite the existence of a reason to accept religion in other cases. I will return to what securitisation means for toleration in the final part of the chapter.

Freedom of religion

Moving now from religious toleration to freedom of religion, the latter can, first of all, be understood in two different ways. Freedom of religion might simply denote the *extent* to which people enjoy non-interference in their religious practice. The fewer constraints, the more freedom (Carter, 1999). Freedom of religion can also denote the *legal protection* of this non-interference by means of rights focussed on religious belief, observance, worship, and practice. Freedom of religion then denotes a liberty right on the side of religious people to hold religious beliefs, and to worship and practice on the basis of those beliefs, together with a claim right obliging others not to interfere in their worship or practice. In the remainder of the discussion, I will rely on this sense of freedom of religion. This is a more specific notion, since much of the freedom from non-interference in the former sense can be secured by other means, e.g. by rights protecting more general freedom of association, or by a culture of toleration, which may not be legally enforced at all. The right to freedom of religion protects religious beliefs and observance *qua* religious.

This is fully compatible with there being limits to freedom of religion. Almost all rights are limited: the protection offered is almost never absolute. One of the only exceptions to this rule is in the case of religious freedom, however, where the freedom to hold religious (and other) beliefs is usually thought to be absolute, e.g. in the ECHR article 9. The remaining aspects of freedom of religion concerning outward actions, i.e. the freedom to manifest religious beliefs in worship, practice, and observance, is limited, e.g. by restrictions that are "in accordance with law" and "necessary in a democratic society" as in ECHR article 9(2). In general, rights that are part of a system of rights cannot protect violations of the rights of others. Therefore, if a system of rights is to avoid inconsistency, there must be limits to most rights.

Therefore, there is really only need to talk of a right to freedom of religion if religious beliefs and practices enjoy *a greater degree of protection* from interference than other kinds of beliefs or practices. Rights protection of religion therefore can mean two things. Either it means that the limits beyond which interference is permissible are wider for at least some forms of religious practice than for other kinds of practices, or it means that the protection of religious practices within these limits is stronger in the sense that it takes more weighty countervailing considerations to outweigh the protection than for non-religious practices.

This means that there is a genuine question about whether freedom of religion is really a distinct right. If freedom of religion is merely an instance of a more general freedom of belief and association, then there is no need for a distinct right. On the other hand, if there really is a distinct right to freedom of religion, then there is need for a justification, which tells a convincing story about what is special about religion so that such special protection is required (Laborde, 2017).

Having noted this philosophical debate, I will set it aside for the present discussion. Whether or not freedom of religion is a distinct right in a philosophical perspective, it is in fact enshrined in all of the most important legal documents and invoked politically. If it is not a philosophically distinct right, then my argument about the effect of securitisation for freedom of religion will merely pertain to the part of some more general right that is conventionally referred to as freedom of religion. This difference will not affect the argument I am about to make about the effects of securitisation, only the correct categorisation of what the affected right is.

Regarding the relation between religious toleration and freedom of religion, there are two different positions on this. One immediately appealing position is that freedom of religion is a way of legally enforcing and securing religious toleration. The state ensures religious toleration in society by granting rights to freedom of religion to people and groups in society who might otherwise be vulnerable to attempts by other groups to interfere with, restrict, or even eradicate their religious beliefs or practices. In a society with rights to freedom of religion, the state will prevent religious intolerance, since competing religious groups will have to refrain from interfering with each other. Call this *the institutionalisation view*, since rights to freedom of religion on this view institutionalise religious toleration.

Against the institutionalisation view, *the pre-emption view* argues that the non-interference secured in such a society is in fact not toleration, since the power condition does not hold. Insofar as the state upholds the freedom of religion, citizens do not have the power to interfere with religious groups they disapprove of – and therefore their non-interference does not count as toleration (Newey, 1999; Heyd, 2008).

One rejoinder to this criticism is that freedom of religion indeed can be a form of toleration, but that we should think of it a part of *a regime of toleration* that secures non-interference and is justified as a means to this (Jones, 2018). This idea is that, even though the individual relations within a society governed by freedom of religion are not relations of religious toleration, the legal framework governing the society can nevertheless be understood in terms of toleration.

Again, I will set this debate aside for present purposes. My argument about the effects of securitisation is independent of whether the institutionalisation or the pre-emption view is the right one. The only difference will concern the conceptualisation of what securitisation has effects *for*.

This is due to some parallels between religious toleration and freedom of religion that are central to the argument I will make. The claim right against interference with religion is somewhat analogous to the acceptance component in religious toleration: both serve to override objections and prevent interference. The difference is that rights are institutional and enforced mechanisms for doing this, whereas the acceptance component denotes reasons operative in the practical deliberation of agents. The other parallel is between the rejection component of religious toleration and the limits of rights protection. In both cases, there is a point beyond which religious practices are not protected or not tolerated, despite the existence of a right to religious freedom or reasons for acceptance.

Just as with religious toleration, there is an immediately evident link between securitisation of religion and limits on rights protection. If a religious practice is perceived as a threat, then it might not be protected, despite the existence of a general right to freedom of religion. The most common example of this is the kind of public order clauses that qualify most rights to freedom of religion.

Toleration and rights as mechanisms for liberal legitimacy

I will now sketch a very general perspective on what the function of religious toleration and freedom of religion is within a liberal conception of state and society. This is not so much a justification, since it will not say anything about what the extent of rights protection should be or about which practices should be tolerated. I will rather sketch a very general framework for understanding what function religious toleration and rights to freedom of religion play within which more specific justificatory questions can be addressed. The argument I will subsequently make about the effects of securitisation pertains to this general function, not to which specific practices to tolerate or protect, or the more specific reasons why. I will therefore not say anything about what the more specific justification for rights is (e.g. whether we should adopt an interest-theory or a will-theory of rights) or what types of reasons are appropriate as reasons for acceptance in relation to toleration (e.g. whether we should adopt a Kantian respect-perspective, as suggested by Forst, 2013).

My suggestion is that freedom of religion and religious toleration serve the general function of securing both political and social *legitimacy*. Precisely because religion is contested, issues of political and social legitimacy arise regarding how state and society relate to religion. Toleration and protection by rights are mechanisms for securing legitimacy.

Political legitimacy is about the acceptability or justification of political power. A political decision is legitimate, not because of its inherent qualities (e.g. whether it is the right decision or a just law), but because it has been taken by a body with the competence to do so and through a proper procedure (Peter, 2016). Legitimacy minimally means that a political authority is permitted to rule. It may also provide citizens with reasons to obey the decisions. Legitimacy is an important property of a political system since there is bound to be disagreement on particular decisions and policies. Legitimacy provides a way of handling disagreement, namely to distinguish between the issue of which decision is the right one and the issue of what makes decisions acceptable, whether one agrees with them or not.

Issues of legitimacy do not only arise in relation to political power in the standard sense (coercively enforced law etc.). There are also power relations in a broader social sense holding between citizens and groups of citizens, e.g. between parties in civil society. Here too it is problematic if one party can impose its contestable conception of the good on others, e.g. by interfering in or attaching costs to specific forms of worship or observance merely because this group happens to disagree with them. This would not only make for conflictual social relations, but it would also be illegitimate in a sense analogous to the one at stake in the assessment of political decisions not living up to the requirements of political legitimacy. So just as a state needs political legitimacy, a society needs social legitimacy.

Issues of legitimacy arise, roughly, when two conditions are present (1) there is a (social or political) power, which can be exercised over people, and (2) people disagree about the reasons on the basis of which this power should be exercised. These two conditions obtain in relation to religion. Categorisation of people in religious terms can make them objects of both political and social power in ways affecting their possibility for worship, practice, and observance. At the same time, people clearly have different, often contradictory, religious beliefs. They even disagree about the value of religion more generally. Religion also involves different patterns of practice and expectations that are often in conflict. So social and political power cannot avoid affecting religion, and there is no agreement about how such power should be exercised. Religion accordingly gives rise to issues of legitimacy. The question is how a state or society should handle religious differences and whether it can do so in a way securing legitimacy.

Both religious toleration as a practice and freedom of religion as a right have a *content-neutral* structure. This means that, within certain limits such as respect for the rights of others, toleration or rights protection is not conditional on specifics about the religious belief or practice in question. Once something is categorised as a religious belief or practice, it is protected, no matter what its more specific content (again, within certain limits, which are not, however, about the content of the belief or practice). Toleration of a practice happens *despite* the disapproval of the practice (disapproval being a content-sensitive judgment), so the reasons for toleration have to be detached from what one thinks about specific practices. Similarly, rights to freedom of religion protect any religious belief or practice irrespective of what the state, the government, or specific officials might think about it. In a classic phrase, a right is 'a right to do wrong' (Waldron, 1981).

It is precisely in virtue of this content-neutral structure that both freedom of religion and religious toleration can be mechanisms for securing political and social legitimacy. Freedom of religion as a general right is a way of securing legitimacy in how a state relates to religion, because it means that the treatment of religion will not be based on what officials or the government at a given time might think about a specific religion or religious practice. A right means that treatment will not depend on such contested single case assessments. Thereby rights avoid the disagreement on the content of religious beliefs, practices etc. Religious toleration is, analogously, a way of securing, not only a lower level of social conflict but also a social form of legitimacy, since it means that parties to social relations do not let their agreement or disagreement with specific religious practices determine how they relate to them.

There are, of course, limits to how neutral freedom of religion and religious toleration can be. First, the very category of religion, which both freedom of religion and religious toleration presuppose, is not given or neutral. The category of religion is selective, since it picks out certain beliefs, practices, identities, forms of tradition and community, rather than others, as 'religious'. This has an inbuilt tendency to exclude alternative forms of categorisation and understanding, and to privilege understandings of practices and conflicts as religious rather than as expressions of other kinds of reasons, interests, and power relations. Furthermore, the category of religion is biased or partisan in the sense that it is informed by certain (Western, Christian, protestant) understandings of religion, which it lifts up as universal. It thereby forces a range of different identities and practices into a framework alien and to some extent unfavourable to them (Shakman Hurd, 2015).

That religion is a non-neutral construct in this sense is an important corrective to many political debates, which tend to take it as a natural and given category. However, for present purposes, this corrective, once noted, can be set aside. My claim here is not that religion is a neutral *category*, but that toleration and rights have a content-neutral *structure*. Toleration and rights both have objects that have to be categorised in order for agents to determine whether rights protection or reasons for acceptance apply to them. Rights and toleration are content-neutral in the sense that the reasons for protection or toleration tell agents to disregard their particular judgments about the object if reasons for protection or acceptance apply to it.

This content-neutral structure is obviously parasitic on the framing of the reasons for protection or toleration. Rights of the kind enshrined in both national and international law take religion to be a category triggering special protection. As I have already noted, there is a question as to whether this is the philosophically correct answer or whether we should rather understand freedom of religion as a mere instance of more general rights to freedom. As it happens, I agree with Cécle Laborde (2017) that we should understand religion as an interpretative concept in Ronald Dworkin's sense (2011). Rather than taking religion to be a fixed semantic category, instances of which require special protection in law, we should instead ask what reasons for protection we have and construct an interpretative, i.e. value-informed, category of religion on that basis. This requires disaggregating the existing category of religion and recognising that there are a number of different reasons for protection and that these apply in some cases of religion in the traditional sense and not in others.

Were this programme of disaggregation carried through, the criticism of the non-neutrality of religion would not be a problem – not because religion would then be a neutral category, which it would not, but because it would link rights protection to an explicitly value-based category that would carry the reasons for protection on its sleeve, as it were.

In the meantime, however, my argument about the effects of securitisation is unaffected by the fact that the existing category of religion is non-neutral and problematic in the noted ways. My argument concerns the difference securitisation makes for toleration and rights protection. Even though the protected and tolerated category is non-neutral, it is still the object of rights protection and toleration. My argument is that securitisation of the category affects this toleration and rights protection. If my argument goes through, these effects obtain whether or not the category of protection or toleration is non-neutral.

Second, both ways of handling religious differences stand in need of justification. Why legitimacy is a good thing, and how rights and toleration should be implemented in practice, depend on value-assumptions that are not neutral, and which might be contested on religious grounds. Furthermore, both rights and toleration are, as noted, limited. Toleration and rights protection only rely on a general categorisation of some practice as religious (albeit, as just noted, this categorisation is itself not neutral). Decisions *not* to protect or tolerate a practice acknowledged as religious (and thereby covered by the content-neutral reasons for acceptance and protection), on the other hand, has to be content-specific; such decisions require assessment of the specific object of toleration or rights protection. Here legitimacy is secured by basing such assessments on uncontroversial standards (i.e. issue where there is less disagreement), but the assessment as such is content-specific.

Effects of the securitisation of religion for religious toleration

Insofar as religion is securitised in the sense that religion is a motivating force, and specific religious groups in particular are increasingly viewed as potential security threats, this can have a variety of effects for religious toleration and freedom of religion. In this section, I will sketch four possible effects, moving from the most obvious ones flowing directly from the definition of religious toleration to some more mediate ones. The former are conceptual in the sense that they relate directly to the definitions of toleration and freedom of religion. The latter rather concern the possibility that securitisation will change the way in which people relate to religion. In that sense, they affect whether religion is an object of toleration or rights protection in the traditional senses at all. Insofar as toleration and rights served the function of securing legitimacy, and insofar as legitimacy is important, such a shift might be normatively problematic.

Starting, then, with the most obvious and immediate effect of securitisation, as already noted, this relates directly to the definition of religious toleration in the sense that concerns for security offer a reason for disapproval or rejection of religion or specific religious beliefs or groups. Since objection is a condition for toleration, securitisation of religion can actually *increase the scope for religious toleration* by increasing objections to religion. Roughly, the more objection to religion, the more room there is for religious toleration. In some cases, however, this will not make a difference if there were already objections to religion. Nevertheless, securitisation makes a difference if it generates objections to religion where none existed before. In these cases, securitisation of religion offers one possible explanation why religion has again become a central issue of toleration.

The second possible effect of securitisation also links directly to one of the conditions defining toleration, namely the rejection component. As also already noted, securitisation of religion can have the effect of shifting the weight of religiously framed threats as possible reasons for rejecting toleration. Even if there is a reason for tolerating religion, this is more easily overridden if religion is viewed as a threat. This means that, whereas the first effect of securitisation implied that the *conceptual room* for religious toleration increased, the *actual extent* of religious toleration is likely to *decrease* as an effect of securitisation. An obvious example of this is the debate about the Islamic headscarf, especially as it has evolved in France, where the issue is not so much the scarves themselves but rather the conflict-seeking groups putatively 'behind' them (Jansen, 2010, p. 78). The *Stasi* Commission's recommendation of a ban on headscarves in public schools, which was subsequently passed in 2004, was motivated by the committee's perception that it was dealing with a security issue that was trumping concern for liberty of religious expression.

Therefore, while the possible scope for religious toleration is wider, the actual extent of religious toleration is likely to be less, since security concerns more often will override reasons for acceptance. This provides one possible explanation for increases in *in*tolerance of religion.

Since there is a parallel between religious toleration and rights to freedom of religion with respect to the rejection component, which is functionally analogous to the limit of rights protection, this possible effect of securitisation also applies to freedom of religion. The concern with security is one standard source of limits to rights since concerns about security translates into concerns for the rights of others. If religion is viewed as a threat, it therefore becomes easier to limit rights protection. Therefore, while the political salience of religion increases the focus on freedom of religion, securitisation might at the same time restrict the actual rights protection of religion.

A third possible effect of securitisation of religion is a variation of the second in that it concerns the rejection component of toleration and the limits of rights protection. As already noted, there usually are limits to both toleration and rights. The ECHR article 9(2) for instance specifies that

> Freedom to manifest one's religion or beliefs shall be subject only to such limitations as are prescribed by law and are necessary in a democratic society in the interests of public safety, for the protection of public order, health or morals, or for the protection of the rights and freedoms of others.

Therefore, any restriction on freedom of religion must be passed by law and justified with reference to one of the noted interests. All of the specified possible legitimate aims of restrictions can relate to considerations of security in different ways. Furthermore, in addition to triggering these limiting considerations directly, securitisation can also affect the *interpretation* of them.

To take an example, in the case of the so-called burqa ban passed in 2010, the French government had to explain why the limitation of freedom of religion constituted by the ban was not a violation of the human right to freedom of religion. The government claimed that the ban was necessary to secure something called "the conditions for living together". It presented this justification as falling under the condition that allows interference with freedom of religion if such interference is justified as a means of protecting the rights and freedoms of others.

The problem with this attempt to establish that the burqa ban did not violate the right to freedom of religion is that there is no such right in the ECHR. Nowhere in the Convention is there any rule that grants citizens a right to see the face of other citizens (Lægaard, 2015b, pp. 210–211). Therefore, the great mystery is why the European Court of Human Rights in the case of *S.A.S. v. France* decided that the burqa ban was in fact a proportionate means to the legitimate aim of protecting the rights and freedoms of others, when there is no such right.

The explanation for this decision is that the Court appealed to the so-called wide margin of appreciation. This is a legal doctrine granting states discretionary power to derogate from rights in the convention. The doctrine was originally developed to handle cases of "war or other public emergency threatening the life of the nation", in which cases the Convention explicitly allows states to set aside rights (Lægaard, 2015b, p. 212). The Court has subsequently extended the doctrine to other cases, including most cases involving religion. Therefore, when the Court appeals to the wide margin of appreciation, it effectively says that cases involving religion can be treated as matters of national security, in which the Court should defer to the judgement of the member states.

At the legal-procedural level, the Court's decision, and the application of the wide margin of appreciation on which it was based, are actually expressions of a general securitisation of religion. Cases involving religion trigger the same discretionary powers on the part of states and the same deferentialism on the part of the Court as national emergencies and security threats do.

What is more, at the more political-substantive level, the reason why the French Government argued for the burqa ban in these terms to begin with is also a form of securitisation. In the explanatory memorandum accompanying and justifying the burqa ban (quoted in sect 25 of the ECHR judgment) the government claimed, among other things, that

> the wearing of the full veil is the sectarian manifestation of a rejection of the values of the Republic. Negating the fact of belonging to society for the persons concerned, the concealment of the face in public places brings with it a symbolic and dehumanising violence, at odds with the social fabric.

To counter this threat, the *Conseil d'État* had

> envisaged an approach based on a new conception of public order, considered in its 'non-material' dimension' and the Government concluded that 'The practice of concealing one's face, which could also represent a danger for public safety in certain situations, thus has no place within French territory.

These quotes exemplify the securitisation of the full veil: even though the government elsewhere admits that the full veil is not in general a security threat and that a general ban therefore cannot be justified on these grounds, it nevertheless continues to mention this as part of a justification of a general ban. Furthermore, it frames the value-based disagreement concerning full-face veils as a matter of security, and even as a matter of countering violence. So even though the justification for the ban is really based on an evaluative disagreement with and criticism of full-face veils, it is systematically framed in terms presenting this as a threat. It is on this basis that the French Government proceeds to present the ban as "necessary in a democratic society" as a means to "protect the rights and freedoms of others". This is a case of securitisation both at the level of the government's framing of the issue and in terms of the explanation for why the Court accepts this otherwise legally unfounded justification.

Here securitisation works in a different way than in the two first possible effects of securitisation: it does not trigger the limiting conditions for rights protection directly, since both the government and the Court admit that full-face veils do not pose a security threat justifying a general ban. Rather, securitisation works, first, by presenting a *different* (value-based) type of justification for the ban, which would not normally be a sufficient legitimate reason for limiting rights protection, *as* a security threat. Second, it further leads the Court to accept this (mis) representation of the justification for the ban, since the Court defers to national governments on religious issues, because they are treated as analogous to issue of national security. The result is that, because religion in general and religious practices like the full-face veil related to Islam in particular are securitised, the right to freedom of religion is limited in ways that it would not otherwise be.

This third possible effect of the securitisation of religion shows that securitisation is a broader process. It affects the understanding of limiting conditions of religious freedom and rejection components of religious toleration by linking them to considerations of security, which allows for the narrower interpretation of freedom of religion and religious toleration.

Securitisation thus affects freedom of religion and religious toleration, not only by directly triggering the established conceptual components but also by shifting their meaning. However, this process can go further than the one already examined. In the case discussed earlier, the framework of freedom of religion as ordinarily understood is still in place, it is only the interpretation of certain limiting conditions that are changed, and only in a specific case. However, securitisation can also affect this framework and the very understanding of freedom of religion and religious toleration that it embodies.

As laid out earlier, freedom of religion and religious toleration share a certain structure, namely that rights protection and toleration of religious practices do not depend on how agents of toleration value these practices. Toleration and rights have a *content-neutral* structure in the sense specified earlier. Furthermore, I have argued that it is in virtue of this structure that rights and toleration can function as *mechanisms of legitimacy*.

A fourth possible effect of the securitisation of religion is to *challenge the content-neutrality* of rights and toleration. This can take place in a roundabout way, which starts out from a liberal self-conception. If we – either as individuals or as a society – conceive of ourselves as liberals, then we are predisposed to view toleration and rights as a positive thing. To be liberal, either as a person or a society, partly means to be tolerant and favour a system of rights protecting individual liberty. We thus link toleration and rights to who we are.

Securitisation works by communicating something as a threat to some referent object. The referent object is usually the citizens or people of a society, e.g. the Dutch, Danes, or Europeans, simply because political communication usually targets the people of a delimited society. If such a people self-identify as liberals, then the liberal identity can be part of what is presented as being under threat. Threats can be presented either as a function of immediate danger, e.g. in the case of terrorism, or as an effect of some foreign intrusion, as in the case of immigration. Both factors are operative in the case of Islam, which is both seen as dangerous and as a foreign imposition. Therefore it is not surprising that almost all European countries have responded to Muslim immigration by articulating 'national values' that, despite the implied particularity of the link to nationality, almost always are universal liberal values such as democracy, equality, freedom (Lægaard, 2007). The result is an assumption of equivalence between the distinction between an external threat and internal referent object, on the one hand, and the distinction between liberal and illiberal, on the other. In other words, it is taken as a given that Europeans are liberal (whether or not they in fact act according to liberal values) and that non-European immigrants, especially Muslims, are illiberal and pose a threat to liberalism.

By thus making liberalism (in this case, toleration and rights) a part of the identity of the referent object, securitisation can change the understanding of liberalism from being a framework for handling legitimate differences of beliefs and values to being a particular set of values and beliefs in competition with others. While liberalism has of course always set limits on religious practices and in that sense has been opposed to certain religious views, the whole point of liberal limits has been to handle religious differences in a legitimate way, rather than taking a stand on which religion is the right one. As already noted, this liberal story can be challenged in a variety of ways, both on the basis that it is more accommodating towards some religious views than others and on the basis that the very category of religion on which it relies is not neutral to begin with. Liberalism (whether of a comprehensive or neutralist, or a more or less secularist variety) is thus not neutral in the big picture (something liberals themselves readily admit). However, as already noted, this does not affect the point that toleration and rights as liberal mechanisms for securing legitimacy have a content-neutral structure. Even if liberalism is not neutral externally, the *raison d'etre* of liberalism is to treat different views equally *within* the liberal framework. Securitisation threatens to undermine this internal aim of liberalism.

To see how this might happen, consider the defining feature of toleration, which is the existence of a space between full acceptance and rejection. Toleration is not relevant when there is no objection, on the one hand, and it stops at some limit marked by the rejection component, on the other. We can only speak of toleration if there is some distance between the point where objection replaces full acceptance, but where the agent of toleration still has reasons for acceptance, and the point where reasons for acceptance give way to rejection. Something similar is the case for rights protection. Rights are only relevant in cases where someone might interfere with some practice or some activity might be vulnerable to interference of domination in the absence of protection. However, rights protection also has limits. The acceptance component and the analogous rationales for rights protection are what keep this space open between full acceptance and full rejection.

When toleration and rights protection become part of the identity of the referent object, which is presented as threatened by (dangerous and foreign) religion, the conceptual components defining this space are easily challenged. If religion is a threat to liberal toleration rather than a legitimate difference to be handled through toleration, it is hard to sustain the space of religious freedom and toleration. Toleration moves from being the framework for handling legitimate religious differences to being itself a competitor to threatening religion. When this happens, the acceptance component that used to keep the space of toleration open by bracketing judgments on practices is transformed to an assessment of whether these practices conform to liberal values. Toleration changes from being content-neutral to a conditional (i.e. content-specific) acceptance of what is valuable. This means that the space between acceptance and rejection that defines toleration in the traditional sense collapses. It is replaced by a dichotomy between practices that are acceptable from a liberal point of view and practices that are not. The former are protected; the latter are rejected. There is no longer any toleration in the classical sense, since the combination of objection and content-neutral acceptance is replaced by content-specific endorsement or rejection.

There are many examples of this kind of development. European states that self-identity as liberal and take pride in having established systems of rights and religious toleration increasingly require religious minorities – and especially Muslim immigrants – to accept the 'national' liberal values. Rather than acceptance of religious differences, this requirement often takes the form that minorities have to hold liberal beliefs themselves. Some naturalisation tests are designed to probe whether applicants really accept liberal beliefs and applicants are sometimes subsequently required to state affirmation of liberal values. Simultaneously with invocations of the value of free speech for liberal legitimacy, states increasingly criminalise expressions of illiberal views. In 2016, the Danish parliament for example passed a law against so-called hate-preachers, which criminalised and otherwise placed restrictions on expressions of anti-democratic views, but only when uttered by religious preachers as part of religious instruction. This is again an example of the securitisation of religion, since this kind of restriction of free speech is only held to be warranted when anti-democratic beliefs are voiced by religious representatives.

When the securitisation of religion works like this, it changes the meaning and function of religious toleration and rights protection. What was earlier a content-neutral mechanism for handling religious differences in a legitimate way turns in to a content-sensitive sorting of religious groups and practices into those that are liberal and hence approved and those that are illiberal and hence rejected. The space for toleration of disapproved practices disappears and is replaced by a comprehensive liberal interventionism and militancy vis-à-vis religion.

Conclusion

Insofar as it is their content-neutral structure that allows the traditional forms of toleration and rights protection to serve as mechanisms of legitimacy, and insofar as one thinks that preservation of legitimacy is an important mark of a liberal society, this development is normatively problematic. In effect, it means that self-identifying liberal societies are undermining their own liberal credentials in the attempt to defend liberal values. I have suggested that this development is primarily one that happens in relation to religion, and especially in relation to religious groups of immigrant origin. From a liberal point of view, however, it is no consolation that this is only happening in some cases – this is in itself a symptom of unequal treatment. Furthermore, I have suggested that the securitisation of specific religious groups has a tendency to spread to religion more generally. This is already evident in as different cases as the European Court of Human Rights' doctrine of the margin of appreciation and the Danish law about hate-preachers. It is even more disturbing if one believes that religious toleration is the original form of liberal toleration more generally, as is historically the case. If this original form of toleration is now undermined, one might worry that this could spread to other forms of toleration as well.

Note

1 Earlier versions of this paper were presented at the conference *Putting Truth in the Second Place: On Compromise, Religion and Politics*, at the University of Copenhagen, and the ACCESS Europe conference *Spaces of Tolerance: The Politics and Geopolitics of Religious Freedom in Europe*, Amsterdam. Thanks to Avi Astor, Carsten Bagge Laustsen, Hans Bruun Dabelsteen, Denis Lacorne, Glen Newey, Noel Parker, Thijl Sunier, and Lorenzo Zucca for helpful comments.

References

Balint, P. (2017). *Respecting toleration: Traditional liberalism and contemporary diversity*. Oxford: Oxford University Press.

Balzacq, T. (2010). *Securitization theory: How security problems emerge and dissolve*. London: Taylor and Francis.

Carter, I. (1999). *A measure of freedom*. Oxford: Oxford University Press.

Cohen, A.J. (2014). *Toleration*. Cambridge: Polity Press.

Dworkin, R. (2011). *Justice for hedgehogs*. Cambridge, MA: Harvard University Press.

Forst, R. (2013). *Toleration in conflict*. Cambridge: Cambridge University Press.

Heyd, D. (2008). Is toleration a political virtue? In: M.S. Williams and J. Waldron, eds., *NOMOS XLVIII: Toleration and its limits*. New York: New York University Press.

Jansen, Y. (2010). Secularism and security: France, Islam, and Europe. In: L.E. Cady and E. Shakman Hurd, eds., *Comparative secularisms in a global age*. New York: Palgrave Macmillan.

Jones, P. (2018). *Essays on toleration*. London: Rowman & Littlefield.

Kaya, A. (2012). Backlash of multiculturalist and republicanist policies of integration in the age of securitization. *Philosophy and Social Criticism*, 38(4–5), pp. 399–411.

Laborde, C. (2008). *Critical republicanism*. Oxford: Oxford University Press.

Laborde, C. (2017). *Liberalism's religion*. Cambridge, MA: Harvard University Press.

Laustsen, C.B. and Wæver, O. (2000). In defence of religion: Sacred referent objects for securitization. *Millenium*, 29(3), pp. 705–739.

Lægaard, S. (2007). Liberal nationalism and the nationalisation of liberal values. *Nations and Nationalism*, 13(1), pp. 37–55.

Lægaard, S. (2013). Danish anti-multiculturalism? The significance of the political framing of diversity. In: P. Kivisto and Ö. Wahlbeck, eds., *Debating multiculturalism in the Nordic welfare states*. Basingstoke: Palgrave Macmillan, 170–196.

Lægaard, S. (2015a). Attitudinal analyses of toleration and respect and the problem of institutional applicability. *European Journal of Philosophy*, 23(4), pp. 1064–1081.

Lægaard, S. (2015b). Burqa ban, freedom of religion and 'living together'. *Human Rights Review*, 16(3), pp. 203–219.

Newey, G. (1999). *Virtue, reason, and toleration: The place of toleration in ethical and political philosophy*. Edinburgh: Edinburgh University Press.

Patten, A. (2014). *Equal recognition*. Princeton, NJ: Princeton University Press.

Peter, F. (2016). Political legitimacy. In: E.N. Zalta, ed., *The Stanford encyclopedia of philosophy*. Summer ed. Available at: http://plato.stanford.edu/archives/sum2016/entries/legitimacy/

Shakman Hurd, E. (2015). *Beyond religious freedom: The new global politics of religion*. Princeton, NJ: Princeton University Press.

Waldron, J. (1981). A right to do wrong. *Ethics*, 92(1), pp. 21–39.

6 Militant secularism versus tolerant pluralism

A critical assessment of the European Court of Human Rights

Margherita Galassini[1]

Europe is witnessing a resurrection of religion and – with it – a re-emergence of old and long-debated controversies concerning the appropriate place of religion in liberal democratic societies. Needless to say, this is a profoundly divisive matter, especially in a time in which European states desperately try to fight the 'invasion' of refugees and migrants – most of whom are Muslim – by building walls, weakening refugee protection, and promoting European values and traditions.

Over the past decades, the European Court of Human Rights (ECHR) has acquired a central role in determining the appropriate space of religion in public life. This has not been an easy task, especially considering the wide range of different government-religion relations that European states exhibit; these "are the result of a variety of historical, social, political and cultural factors [and – as such – they] should in principle be respected" (Martínez-Torrón, 2012a, p. 331). Hence, the ECHR has attempted to achieve a balance between diversity and universality in the protection of religious freedom across Europe by appealing to the doctrine of the 'margin of appreciation' (Martínez-Torrón, 2012a, p. 332), which is "an international law doctrine of judicial self-restrain or deference" (Lewis, 2007, p. 397). The key idea is that even though liberal democracy requires some degree of separation between state and religion, there is no unique formula that can – or should – be applied to all countries.

The ECHR has been accused of promoting a form of 'militant secularism', which authorises a state to act in a militant manner – by imposing strict restrictions on the exercise of religious freedom – in order to safeguard its democracy's core constitutional commitment to secularism (Macklem, 2012). The aim of this chapter is twofold. First, it urges the ECHR to adopt an approach to religion similar to Laborde's minimal secularism, which is designed to set out the necessary conditions for liberal legitimacy, while allowing states to pursue their own reasonable interpretation of the principles of justice (Laborde, 2017). This framework seems to be particularly well suited for the European context, where the ECHR's doctrine of the 'margin of appreciation' may be regarded as the tool that allows for the creation of a European space of tolerance understood figuratively – that is, as the realm of possibility within the domain of reasonable liberal conceptions of justice.

Second, it assesses whether the ECHR's protection of the right to religious freedom is sufficient at the bar of liberal legitimacy by discussing some its most debated rulings on the presence of religious symbols in the public sphere. This exercise allows me to highlight some points of disagreements with Laborde's approach and suggest a revised version of her account that I term tolerant pluralism. Broadly speaking, it is my contention that Laborde's notion of minimal secularism is still too secularist with regard to state officials' right to wear – and states' permissibility to display – religious symbols in the public sphere.

This chapter proceeds as follows: I begin with briefly explaining the ECHR's approach to religious freedom and the notion of minimal secularism that Laborde develops in *Liberalism's Religion*. I then discuss cases concerning the permissibility of wearing personal religious symbols in the public sphere and assess them against the value of liberty entailed in minimal secularism.[2] I first discuss the headscarf 'affair' in Turkey and in France, where the prohibition on wearing the Islamic veil applies to ordinary citizens and argue that the ECHR's ruling is inconsistent with liberal legitimacy as it justifies restrictions on the right to religious freedom by appeal to a comprehensive notion of secularism. Indeed, the ECHR's treatment of religion in the headscarf affair may be described as an instance of militant secularism. I then consider another instance of the headscarf affair – this time in Switzerland – where the prohibition targeted a state official. I claim that in the Swiss case, the ruling illegitimately violated the applicant's personal liberty. Last, I discuss the permissibility of displaying institutional religious symbols in the public sphere by focussing on the crucifix case in Italy; here, I defend the ECHR's judgment and disagree with Laborde's view that it violates the third condition of minimal secularism – namely civic equality.

The ECHR and minimal secularism

Unlike some domestic constitutions, the European Convention on Human Rights does not mandate a particular degree of separation between religious and political institutions, nor does it proscribe establishment (Evans and Thomas, 2006, p. 699). Indeed, the signatory states to the Convention exhibit significant differences in their specific relationships between church and state, with some countries upholding a strict principle of secularism (e.g. Turkey and France) and others supporting an established church (e.g. United Kingdom, Denmark, and Greece). As Evans and Thomas point out, however, the ECHR does "indirectly regulate the permissible forms of relationship between religious institutions and the state by reference to religious freedom" (Evans and Thomas, 2006, p. 699).

The European Convention on Human Rights contains three provisions that deal with religion. Article 9 sets out the basic framework for the protection of the right to religious freedom:

1   everyone has the right to freedom of thought, conscience and religion; this right includes freedom to change his religion or belief and freedom, either alone or in community with others and in public or private, to manifest his or her religion or belief, in worship, teaching, practice, and observance

2   freedom to manifest one's religion or beliefs shall be subject only to such limitations as are prescribed by law and are necessary in a democratic society in the interests of public safety, for the protection of public order, health or morals, or for the protection of the rights and freedoms of others (Council of Europe, 1950)

Article 14 ensures that the enjoyment of the rights laid down in the Convention shall be free from religious discrimination; finally, Article 2 of the first Protocol to the Convention requires that "the State shall respect the right of parents to ensure [. . .] education and teaching in conformity with their own religious and philosophical convictions" (Council of Europe, 1950).

   Article 9 is the central provision regulating the right to freedom of religion; it is this Article that this chapter focusses on. Religious freedom has a twofold dimension: the *forum internum* (the internal dimension), which entails the freedom to *hold* one's religion or belief, and the *forum externum* (the external dimension), which protects the freedom to *manifest* one's religion or belief either in public or private (Martìnez-Torrón, 2012b, p. 367). The former dimension of religion is absolute and inviolable. It is only the latter dimension that may be subject to limitations as indicated under the second paragraph of the Article. The rationale behind paragraph 9.2 lies in the assumption that the conduct of a religious believer who publicly manifests his or her religion will be more likely to negatively affect others than the conduct of someone whom holds his or her belief within the private domain (Lewis, 2007, p. 400). At the same time, an area of uncertainty exists because the precise meaning of 'public sphere' was never specified (Ringelheim, 2012, p. 285).

   The doctrine of the 'margin of appreciation' is the main instrument used by the ECHR to achieve a balance between diversity and universality – that is, between the respect for national church-state relations and the commitment to offer an equal degree of protection of religious freedom across Europe (Martínez-Torrón, 2012a, p. 331–332). The ECHR tends to employ a wide margin in areas where there is a lack of consensus or common practices across Europe – such as in the fields of morals and religion (Lewis, 2007, p. 397). As the ECHR has repeatedly pointed out, "it is not possible to discern throughout Europe a uniform conception of the significance of religion in society [. . .]; even within a single country such conceptions may vary" (*Otto-Preminger-Institut v. Austria*, 1994, p. 43). Consequently, "where questions concerning the relationship between State and religions are at stake", the ECHR grants national authorities a degree of flexibility as to how they protect the right to religious freedom (*Leyla Sahin v. Turkey*, 2005: para 109); for the domestic context strongly determines "the meaning or impact of the public expression of a religious belief" and thus "the requirements imposed by the need to protect the rights and freedoms of others and to maintain public order" will vary from state to state (*Leyla Sahin v. Turkey*, 2005: para 109). National authorities are therefore in principle considered to be in a better position than the ECHR to assess whether restrictions on religious freedom are necessary or justifiable.

   Political liberalism is committed, at least minimally, to some degree of separation between the state and religion. In *Liberalism's Religion*, Laborde defends her notion of minimal secularism, which she conceives as "a minimal normative requirement of liberal legitimacy [. . .], a political doctrine specifying the rightful place of religion in the state" (Laborde, 2017, p. 113). The chief argument of Laborde's liberal egalitarianism is that the notion of religion should be interpreted and disaggregated into its multiple dimensions (Laborde, 2017, p. 2). Framing the issue of religion in terms of the values that it realises has the straightforward implication of making religious and non-religious beliefs analogous. That is to say, whatever protection or containment is appropriate with regards to religion, it is so because of some feature that it shares with non-religious beliefs (Laborde, 2017, p. 3).

Minimal secularism appears as a central component of Laborde's liberal egalitarian account of religion and the state. Contrary to existing liberal egalitarian theories (Dworkin, 2013; Eisgruber and Sager, 2007; Taylor and Maclure, 2011; Schwartzman, 2012, 2017; Quong, 2011), Laborde seeks to articulate an understanding of the secular state that goes beyond the ideal of state neutrality toward the good. Disaggregating religion thus allows us to identify the three dimensions of religion that directly affect liberal legitimacy. The liberal state may not endorse or recognise a religious belief or practice that is inaccessible, comprehensive, or divisive. In turn, these three features of religion relate to the three liberal values that constitute minimal secularism: epistemic accessibility (the justifiable state), liberty (the limited state) and civic equality (the inclusive state) (Laborde, 2017, p. 117).

The strength of Laborde's approach consists in her acknowledgement that liberal egalitarianism is consistent with a wider variety of state-religion relations than liberals have commonly assumed; that is, according to the normative standards of liberal egalitarianism, many different models of secularism – or separation between state and religion – are legitimate (Laborde, 2017, p. 116). The crucial point here is that, in her view, political liberalism allows for some degree of inconclusiveness with regard to the interpretation of the principles of justice. Hence, once the three liberal values of minimal secularism are set out – accessibility, personal liberty, and civic equality – there is still room for reasonable disagreement about how these ideals are to be interpreted.

As I argue throughout the chapter, it is my contention that Laborde's approach is still too secularist with regard to the presence of religious symbols in the public sphere. Consequently, I defend a revised account – which I term tolerant pluralism – aimed at according wider protection to religious manifestations; particularly, as I show in my discussion of *Dahlab* and *Lautsi*, I am in favour of a more permissive stance on state officials' right to wear and states' permissibility to display religious symbols in the public sphere. While maintaining the core structure of Laborde's minimal secularism – with its three conditions for liberal legitimacy – my account allows for a more tolerant balancing of the interests and values at stake when assessing the permissibility of religious manifestations in public.

The language of toleration and pluralism – rather than secularism – is closer to Rawls' understanding of the relation between religion and democracy as well as to the lexicon used by the ECHR. Indeed, Rawls never uses the term 'secularism', but rather appeals to the ideals of toleration or – more rarely – neutrality. In her contribution to this volume, Gentile stresses how – according to Rawls – tolerance is first and foremost a *political* value, which is built on the overlapping consensus between incommensurable comprehensive doctrines and is committed to a "neutrality of aim" whereby "liberal institutions must secure the equality of opportunities to pursue any 'permissible' worldview" (Gentile in the current volume); indeed, "it would be at odds with tolerance to suppress religious practices and ways of life" (Gentile in the current volume). In this way, the notion of tolerant pluralism captures the ECHR's role of impartial protector and organiser of religious pluralism.

Further, it is possible to discern some principles that have progressively emerged in the ECHR's case law on religious freedom; these closely mirror the principles underpinning the right to religious freedom that have been prominently defended in the liberal tradition, and they are pluralism, autonomy of religious communities and state neutrality. In sum, I endorse an approach which is built on the framework provided by Laborde's minimal secularism and yet allows for greater variety in state-religion relations, while using a language that is more faithful to the Rawlsian tradition and the ECHR's increasing effort at "building a theory" of religious freedom that is valid across Europe (Ringelheim, 2012, p. 284).

The role of the ECHR can thus be conceived as that of ensuring that states meet the three conditions of liberal legitimacy as entailed by Laborde's minimal secularism, while allowing them to pursue their own reasonable interpretation of such ideals. The ECHR's doctrine of the 'margin of appreciation' can be regarded as the tool that allows for the creation of a European space of tolerance understood figuratively – that is, as the realm of possibility within the domain of reasonable liberal conceptions of justice.[3] In other words, the ECHR should endorse a notion of tolerant pluralism so as to be able to accommodate, by appealing to the 'margin of appreciation' doctrine, the national variations in the reasonable interpretations of the principles of justice across Europe.

The rest of chapter focuses on the liberty and equality conditions, for the requirement of accessibility – which Laborde conceives as "public reason *stricto sensu*" (Laborde, 2017, p. 119) – "is a necessary but not a sufficient condition for the law's liberal quality" (Laborde, 2017, p. 123). Indeed, "the accessibility condition is intended not as a final test but as a prior test of permissibility. It [. . .] does not specify which reasons are conclusive enough to provide a full justification for public policy and law" (Laborde, 2017, p. 130). As I show in the next sections, the ECHR's defence of the headscarf bans and of the display of the crucifix in public schools relies on public reasons; the problematic aspects of the ECHR's rulings, however, concern an all-things-considered judgment about the values of liberty and equality.

Personal religious symbols and the limited state

Political liberalism is strongly committed to the value of personal liberty. Citizens' ability to develop as self-determining agents prevents the state from enforcing comprehensive doctrines that would interfere with each individual's right to form judgments about how to lead their lives (Laborde, 2017, p. 144). Laborde suggests two ways in which a coercive law violates personal liberty: (1) if the law is justified by appeal to a comprehensive conception of the good; (2) if, regardless of the law's justification, it limits one's liberty to live with integrity (Laborde, 2017, p. 146). Before turning to the discussion of the headscarf cases, I should explain why Laborde distinguishes between conditions (1) and (2).

In her discussion of state neutrality, Laborde distinguishes between the scope (broad or restricted) and the focus of neutrality (Laborde, 2017). The latter implies that both the justification as well as the subject-matter of a policy ought to be neutral. "At the level of justification, neutrality applies to comprehensive conceptions of the good" – i.e. the state ought not to justify a policy by appeal to a comprehensive worldview – condition (1) (Laborde, 2017, p. 146). At the level of the subject-matter, on the other hand, neutrality applies to "integrity-related liberties" – condition (2) (Laborde, 2017, p. 146). Indeed Laborde claims that, first, a law may be justified by a comprehensive conception of the good without impinging on salient freedoms – for instance if the state implements restrictions of free movements on Shabbat[4]; this law is impermissible if it is justified on the grounds that it allows "citizens to live virtuously, according to an authentic and true conception of the good life" (Laborde, 2017, p. 146). Second, a law may be justified by "partial ideas about the good (such as the badness of addiction, or attachment to tradition) and public goods (social cohesion, public health, rules of social coordination, and so forth)" and yet be impermissible because it infringes upon an integrity-related interest (Laborde, 2017, pp. 146, 147). The distinction between conditions (1) and (2) is crucial in order to understand the difference between the cases to which I now turn.

Religious symbols and ordinary citizens

In *Leyla Sahin v. Turkey*, the applicant is a female medical student at Istanbul University who was prohibited from wearing the Islamic headscarf within university facilities. The ECHR upheld the Turkish Constitutional Court's decision to ban headscarves and other forms of religious dress in Turkish universities. The ban was motivated by the necessity to protect secularism – a fundamental pillar of Turkish democracy. Indeed, Turkey has an explicit constitutional commitment to secularism as indicated by Article 2, which describes it as a "republican, democratic, secular and social state", and Article 4, which ensures the irrevocability of these basic principles (Evans and Thomas, 2006, p. 705).

In its reasoning, the ECHR declares that the headscarves ban pursued the legitimate aims of protecting the rights and freedom of others and securing public order. The ECHR granted Turkish national authorities a wide margin of appreciation as it sought to respect the delicate domestic relations between state and religion. Consequently, the ECHR justified restrictions on individuals' right to freedom of religion as "necessary in a democratic society" in the meaning of Article 9.2.

With regard to the protection of public order, the ECHR stresses that the headscarf ban is necessary to ensure the "peaceful coexistence between students of various faiths" (*Leyla Sahin v. Turkey*, 2005: para 111). This, however, seems to be an unwarranted assumption for – as pointed out in the dissenting opinion of Judge Tulkens – there was no "disruption in teaching or in everyday life at the university, or any disorderly conduct, as a result of the applicant's wearing the headscarf" (*Leyla Sahin v. Turkey*, 2005: para 8).

With regard to the protections of the rights and freedom of others, the ECHR agrees with Turkish authorities that the Islamic headscarf is a "powerful external symbol" which may exert pressure on those who choose not to wear it (*Leyla Sahin v. Turkey*, 2005: para 111). The ECHR recognises that "this religious symbol has taken on political significance in Turkey in recent years". This was seen as particularly worrisome due to the presence of "extremist political movements in Turkey which seek to impose on society as a whole their religious symbols and conception of a society founded on religious precepts" (*Leyla Sahin v. Turkey*, 2005: para 115). Yet, as the dissenting opinion of Judge Tulkens points out, there was no evidence to assume that the applicant wore the headscarf in an "ostentatious or aggressive" manner, or that it was used to "provoke a reaction, to proselytise or to spread propaganda and [. . .] undermine [. . .] the convictions of other[s]" (*Leyla Sahin v. Turkey*, 2005: para 8). Further, Judge Talkens recognises that "merely wearing the headscarf cannot be associated with fundamentalism [. . .]. Not all women who wear the headscarf are fundamentalists and there is nothing to suggest that the applicant held fundamentalist views" (*Leyla Sahin v. Turkey*, 2005: para 10).

The ECHR further declares the headscarf ban necessary in order to protect the rights of women and ensure gender equality, which is one of the core principles underlying the Convention. The dissenting opinion, however, questions the connection between the ban and sexual equality. It points out that "wearing the headscarf has no single meaning [. . .]. It does not necessarily symbolise the submission of women to men and there are those who maintain that, in certain cases, it can even be a means of emancipating women" (*Leyla Sahin v. Turkey*, 2005: para 11). Most importantly, the ECHR fails to take into serious consideration the perspective of the applicant who affirms that she was wearing the headscarf of her own free will.

The ECHR's ruling in Sahin has a significant impact on the way in which the headscarves controversies were dealt with in other European states. In *Dogru v. France*, the applicant was an 11-year-old girl who was expelled from her secondary school because she refused to take off her Islamic headscarf during physical education classes (*Dogru v. France*, 2008).[5] As in *Sahin*, the ECHR does not find a violation of the applicant's right to freedom of religion. Indeed, the European Court explicitly follows the doctrine endorsed in *Sahin*, thus acknowledging that "in France, as in Turkey [. . .], secularism is a constitutional principle, and a founding principle of the Republic, to which the entire population adheres and the protection of which appears to be of primary importance" (*Dogru v. France*, 2008: para 72). In 2004, France passed a law on religious symbols in public schools (but not universities) whereby it banned "the wearing of signs or clothes through which pupils ostensibly express a religious allegiance" (quoted in Laborde, 2008, p. 52).

As in Sahin, restrictions on individuals' right to religious freedom are considered to be necessary to preserve the secularity of public schools, which serves the purpose of protecting the rights and freedom of others and maintaining public order. Once again, however, the ECHR fails to provide sufficient reason to convince that the wearing of the Islamic headscarf does in fact undermine the protection of the rights and freedom of others and the protection of public order. As a result, the restrictions on individuals' right to freedom of religion upheld by the European Court cannot be regarded as "necessary in a democratic society" in the meaning of Article 9.2.

The Turkish and French governments explicitly justify the bans by appeal to the principle of secularism, which the ECHR considers to be consistent with the values underpinning the Convention and necessary to protect the democratic system in these two countries (*Leyla Sahin v. Turkey*, 2005: para 114; *Dogru v. France*, 2008: para 69, 72). Yet, as I have previously stressed, the legitimate aims that the bans are supposed to protect were not in fact jeopardised by Muslim students wearing the headscarf. Consequently, the ECHR unreflectively accepts the Turkish and French conception of secularism as legitimate justifications for the headscarf bans for the purposes of protecting democracy.

It has been argued that the ECHR's judgments in the field of religious freedom have recently begun to treat religion as a threat to the democratic state. Macklem (2012) stresses the principle of militant democracy, which was originally designed to protect the democratic process of a political community, is now advocated in order to defend substantive conceptions of democracy from the influence of religion.[6] In particular, the perception of religious fundamentalism as a new form of threat to the democratic state has incentivised states to view religion with increasing fear. Consequently, while originally conceived as a means for preventing anti-democratic parties from gaining power in the political arena, militant democracy has recently begun to broaden its range of targets. In sum, Macklem suggests that the ECHR has begun to uphold what he refers to as "militant secularism", which authorises a state to act in a militant manner in order to safeguard its democracy's core constitutional commitment to secularism.[7]

According to Laborde, minimal secularism is a normative requirement of liberal legitimacy; hence, ensuring that the three conditions of minimal secularism are met is not only permissible but mandatory for the protection of liberal democracy. Yet, two remarks have to be made here: first, as Laborde's approach amply shows, secularism "should not be reduced to one value, but explicated in relation to a constellation of liberal values" – namely, justifiability, equality and liberty. (Laborde, 2017, p. 115). Second, the three conditions of minimal secularism apply to the state and not to the citizens. Indeed, as Laborde puts it: "the state should be secular so that citizens do not have to be secular" (Laborde, 2013b, p. 169). Hence, secularism is a political doctrine that specifies the obligations of state institutions and officials *towards* citizens. In sum, tolerant pluralism – just like Laborde's minimal secularism – *may* require restrictions on state officials' – but not ordinary citizens' – right to religious freedom for the purpose of protecting the legitimacy of the democratic state. Yet, as I will argue in my discussion of *Dahlab v. Switzerland*, tolerant pluralism is more permissive when it comes to accommodating state officials' right to manifest their religions in the public sphere.

What are we to make then of the ECHR's judgments on the headscarf *affaires* in Turkey and France? It is my contention that the bans violate the third condition of minimal secularism – namely, personal liberty – on the grounds that their justification appeals to a comprehensive conception of secularism. As Plesner suggests, this notion of secularism is "fundamentalist" as it imposes a secularist ideology and way of life on all citizens when they enter the public sphere (Plesner, 2005, p. 3). As a result, secularism becomes the 'religion' endorsed and promoted by the state.[8] I will now illustrate this point further.

Turkey and France endorse a distinctively republican understanding of secularism – that is, *laïcité* – which requires a neutral public sphere in order to secure equal religious rights for all. As Laborde explains, "republican *laïcité* endorses a more expansive conception of the public sphere than political liberalism, as well as a thicker construal of the "public selves" which make up the citizens of the republic" (Laborde, 2005, p. 307). In order to understand this claim, it is necessary to consider what Taylor refers to as "the independent ethic mode of secularism" (Taylor, 1998, p. 33).

Rawls' political liberalism is committed to build a political community grounded on political values that all citizens can agree upon, regardless of their particular comprehensive conceptions of the good. In order to realise this objective, Taylor suggests that, historically, two distinct strategies – or 'modes of secularism' – have been employed (Taylor, 1998, p. 33). The first is the 'common ground' strategy, which sought to identify those doctrines that were common to all theists. The second is the 'independent ethic' approach, which required citizens to abstract from their religious views altogether and identify an independent ground for political morality (Taylor, 1998, p. 33). With the identification of a common ground of convergence, the independent ethic approach was able to promote a strong sense of common identity where it did not previously exist (Taylor, 1998, p. 44).

In the public sphere religion was irrelevant, for it was the independent ethic which reigned sovereign. In the private dimension, on the other hand, the faithful were free to obey to the religious commands of God. As a result, it was necessary to strengthen individuals' identity as citizens so that it would take precedence over other poles of identities, such as class, gender, and, especially, religion; this may occur through the promotion of an express ideology, as in the case of French Republicanism (Taylor, 1998). Indeed, French *laïcité* promotes a robust republican identity grounded on the values of democratic and egalitarian citizenship; in this way the French Republic seeks to transform 'believers' into 'citizens' (Laborde, 2005, p. 316). Republican *laïcité* is therefore the clearest expression of "the independent ethic mode of secularism" (Bhargava, 1998, p. 17).

The Republic has thus to engage in a strong formative project that can allow citizens to develop a robust public identity grounded on the public values of democratic and egalitarian citizenship (Laborde, 2005, p. 317). This imposes strict restrictions on citizens' right to manifest their religion in public: individuals' attempts to express confessional allegiances are easily perceived as signs of proselytism, which infringes the requirement of state neutrality. Schools are especially important as they are "miniature communities of citizens", where students learn the values of public citizenship (Laborde, 2005, p. 327). Understood in this light then, the Islamic headscarf is perceived as an illegitimate act of religious propaganda, which ought to be excluded from the public sphere (Laborde, 2005, p. 327). The headscarf ban is thus understood as a "non-monetary tax imposed on Muslims for the maintenance of the secular state" (McKinnon, 2007, p. 137).

Two points have to be made here. First, it now appears clear that the headscarf bans were motivated by a commitment to a comprehensive notion of *laïcité*, which sought to marginalise the role of religion so as to reduce its influence in social life (Laborde, 2013b, p. 183). Hence, Plesner is right in arguing that the ECHR accepted a notion of 'fundamentalist secularism' which violates individuals' personal liberty and is thus impermissible at the bar of liberal legitimacy. Second, republican *laïcité* ought not to be rejected as necessarily inconsistent with the requirements of minimal secularism and liberal legitimacy, for, when properly understood, it is compatible with the wearing of religious signs by pupils.

Indeed, in defence of a what she coins 'critical republicanism', Laborde presents a more nuanced interpretation of the values promoted by *laïcité* and suggests that the demands of secular neutrality ought not to apply to ordinary citizens – in these cases, schoolchildren and university students. While it was true from the start that the school itself had to be subject to a requirement of neutrality, this was only "meant to apply to teachers, the content of teaching and school building, not to pupils themselves" (Laborde, 2008, p. 60). For, as Laborde succinctly puts it: "the demand, made in the name of French *laïcité*, that state school pupils, or users of public services, show restraint in the expression of their religious beliefs is an illegitimate extension of the demands of secularism from the state to citizens"(Laborde, 2013b, p. 169). In sum, the suitably amended notion of *laïcité* is fundamentally committed to religious freedom, civic inclusion, and fairness; when suitably interpreted, it is not inconsistent with the wearing of religious symbols in the public sphere by ordinary citizens.

Religious symbols and agents of the state

In *Dahlab v. Switzerland*, the applicant was a primary school teacher who was prohibited from wearing the Islamic headscarf when she was performing her professional duties (Dahlab v. Switzerland, 2001). The applicant appealed to the ECHR and complained of an infringement of her right to manifest her religion.

The ECHR considers that the measure prohibiting the applicant from wearing a headscarf while teaching was "necessary in a democratic society" for the protection of the rights and freedoms of others, public order, and public safety. Indeed, the ECHR notes that the Federal Court justifies the prohibition on the grounds that teachers are under a strict obligation to respect the principle of denominational neutrality in state schools, which ensures the protection of the religious beliefs of pupils and their parents and the promotion of tolerance and religious harmony. In particular, as civil servants, teachers represent the state; hence, their conduct should not suggest that the state sides with one religion rather than another (*Dahlab v. Switzerland*, 2001). Further, the ECHR acknowledges that the Islamic headscarf is a "powerful external symbol" whose proselytising effect may influence the applicant's pupils given their tender age (they were between four and eight years old). Consequently, after weighting the applicant's right to manifest her religion against the need to ensure religious harmony in schools and protect pupils' and parents' religious beliefs, the ECHR concludes that it was within Switzerland's margin of appreciation to prohibit the wearing of the veil to teachers in state schools and hence finds no violation of the applicant's right to religious freedom.

Yet, doubts may be cast on whether the prohibition of the wearing of the Islamic headscarf was indeed necessary for the protection of the legitimate aims recognised by the ECHR. For, as the applicant argues, the fact that the she wore an Islamic headscarf had gone unnoticed for years and no complaints had been made by pupils or parents during a period of more than five years. Further, the applicant had always experienced towards her and, as a result, religious harmony had never been disturbed within the school (*Dahlab v. Switzerland*, 2001, p. 11). Hence, this seems to imply that the fact that applicant wore the headscarf did not cause any disturbance in her pupils. In sum, just as in *Sahin* and *Dogru*, there seems to be no evidence in support of the ECHR's conclusion that the violation of the applicant's right to manifest her religion is necessary within the meaning of Article 9.2

It is thus plausible to question whether the prohibition was once again motivated by a comprehensive notion of secularism. Indeed, the principle of denominational neutrality advanced by Swiss national authorities and supported by the ECHR stems from the secular nature of the state and its separation from religion. Further, just like Turkey and France, Switzerland endorses the notion of republican *laïcité*, which, as I have discussed in the previous section, has been interpreted in ways that justify illegitimate violations of religious freedom. Contrary to the previous cases, however, the principle of denominational neutrality applies to state institutions and representatives, rather than to users of public services – such as pupils. As Laborde suggests, the permissible prohibition of religious sign should depend on the public function of the state official in question and on the vulnerability of the users of the service (Laborde, 2008, p. 87). Consequently, "government ministers but not tax inspectors, primary school teacher but not university lecturers, may be subjected to an obligation of religious restraint while on duty" (Laborde, 2008, p. 87).

Nonetheless – contrary to Laborde – it is my contention that in the case at hand the prohibition violates the third condition of minimal secularism, namely personal liberty, for it limits the applicant's liberty to live with integrity. Indeed, as Laborde herself suggests, if a law "is justified by sound neutral reasons, but nonetheless burdens a central Muslim practice, it "might not be all-things-considered justified if it gravely infringes a salient interest" (Laborde, 2017, p. 201). Laborde argues that Muslim women have an integrity-related claim to wear a headscarf. Consequently, Muslim veiling "cannot be simply trumped by any appeal to the general welfare, public order, and so forth" (Laborde, 2017, p. 199). This idea is grounded on Rawls' thin theory of the good, according to which certain rights and freedoms are considered to be essential to the exercise of basic human capacities and hence acquire priority over other freedoms or aims. These essential freedoms include conscience, speech, and association.

Integrity entails coherence between one's actions and one's ethical commitments so that if one is forced to act in a way that is contrary to his or her ethical commitments, he or she will inevitably feel shame, guilt, and remorse (Laborde, 2017, pp. 203, 204). Laborde refers to these as integrity-protecting commitments (IPCs): "commitments, manifested in practice, ritual, or action (or refusal to act), that allow an individual to live in accordance with how she thinks she ought to live" (Laborde, 2017, p. 203). The integrity-protecting commitment must be non-trivial and important for the individual (thick sincerity test). Further, it must not be morally abhorrent (thin acceptability test). Muslim veiling can thus be regarded as an instance of obligation-IPCs – which Laborde distinguishes from the less stringent identity-IPCs (Laborde, 2017, p. 215) – since most women experience it as an obligation. On Laborde's interpretation of freedom of religion, what matters is not whether veiling is in fact a core practice in the Islamic religion, but rather whether women perceive it as a practice that directly affects their integrity; as such, legal burdens upon them are particularly severe (Laborde, 2017, p. 223).

Obligation-IPCs should be protected against disproportionate burden. Laborde puts forward four criteria to assess whether a burden is disproportionate in relation to the aims pursued by the law (Laborde, 2017, p. 221). Let us see how these criteria are fulfilled in *Dahlab*.

1   *Directness*. "The directness of a burden is measured in relation to the costs incurred by individuals in avoiding subjection to the law or regulation in the first place" (Laborde, 2017, p. 221). The prohibition of the wearing of the Islamic headscarf in the case at hand was directly burdensome. The applicant did not have the choice to teach in a private school as many were not in the Canton of Geneva and, furthermore, they were not of Muslim faith. Failure to comply with the school's directives would de facto prevent her from working as a teacher.
2   *Severity*. "The more an IPC is perceived as an obligation, the more severe the burden is" (Laborde, 2017, p. 225). Muslims women generally experience the wearing of the Islamic headscarf as a religious obligation. This was certainly so in the case of *Dahlab*.
3   *Aim of the law.* "The more tightly a law promotes a goal of egalitarian justice and the more it requires a universal and uniform compliance for its effectiveness", the harder it will be for the claimant to be alleviated from the law's burden (Laborde, 2017, p. 225). The purpose of the law is egalitarian as it sought to protect children's right to religious freedom and religious harmony.

4   *Cost shifting*. The fact that she wore the headscarf did not result in any disturbance in the school. Hence, the applicant's failure to comply with the prohibition would not result in any cost shifting.

In sum, in the present case, the balance of these reasons renders the burden disproportionate. Hence, I object to the ECHR's ruling in *Dahlab* on the grounds that the
prohibition to wear the Islamic headscarf resulted in a violation of the applicant's
personal liberty. According to my understanding of tolerant pluralism, the prohibition is impermissible at the bar of liberal legitimacy.[9]

Institutional symbols and civic equality

In *Lautsi v. Italy* the applicant is the parent of two children who attended a state
school that displayed a crucifix in each classroom (*Lautsi v. Italy, 2011*). After
the husband's applicant raised the issue of the permissibility of the presence of
crucifixes in classrooms, the question was put to a vote and the majority of the
school's governors decided to keep the religious symbol in place. The applicant
brought the case before the national authorities without success. The applicant
then complained before the ECHR arguing that the display of the crucifix in state
schools illegitimately violated her children's right to religious freedom (Article
9) and education (Article 1 of Protocol No. 1). However, The ECHR found no
violation of Article 9 and 1 Protocol No. 1 and concluded that the issue of the
presence of religious symbols in classrooms is a matter falling within the margin
of appreciation of the state.[10]

It is possible to identify two main arguments advanced by the Italian government in support of the display of the crucifix: I will refer to them as the
liberal and as the cultural argument respectively.[11] The former emphasises the
liberal content of European Christianity, which, in its recent history, has helped
promote the liberal values of "tolerance, mutual respect, valorization of the
person, affirmation of one's rights" (*Lautsi v. Italy, 2011*: para 16). As a result,
the crucifix becomes the symbol of equal inclusion, rather than an implicit
message of sectarianism. In the words of the Italian government, "the cross, as
the symbol of Christianity, can therefore not exclude anyone without denying
itself" (*Lautsi v. Italy, 2011*: para 15). The crucifix may thus acquire an educational symbolic function as it affirms – rather than undermines – the values
underpinning the Italian constitutional order and the principles of secularism
itself.

Yet this line of argument encounters two criticisms: first, there might be
more effective ways of promoting the liberal values at the foundation of Italian society and its constitutional order, for instance through the presence in
the school curriculum of a subject that teaches the values and history underpinning the Italian political order. Second, it might be difficult for people
who are not Christian to find any educational value in the crucifix, since its
meanings might not be accessible to them – precisely because of its Christian
origins.

The latter argument advanced by the Italian government detaches the crucifix from its theological origins and instead focusses on the cultural dimension that it has acquired through history, thus becoming a secularised symbol of the Italian national identity. Indeed, the Italian government points out that "keeping crucifixes in school was therefore a matter of preserving a centuries-old tradition" (*Lautsi v. Italy, 2011*: para 36).

According to Laborde, however, it is not sufficient to deprive the symbol of its religious significance in order to make it permissible at the bar of liberal legitimacy. Indeed, Laborde explicitly disagrees with the ECHR's decision in *Lautsi* because, according to the expressionist theory of non-endorsement that she defends, religious symbols may have exclusionary valence; that is, if a religious symbol has a social meaning which may cause social discrimination, then the state ought not to display it (Laborde, 2017, pp. 137–140). Laborde explicitly mentions the crucifix in schools as an example of impermissible display of a religious symbol by the state. Throughout history religion has been "a socially salient category of membership and exclusion" in European societies (Laborde, 2017, p. 137). As *Lautsi* evidently shows, European states have begun to re-describe the crucifix as a cultural rather than religious symbol; yet, their attempt to secularise the meaning of the crucifix and to ascribe it to the cultural and national identity does not make it less problematic for "both faith and culture can be used as markers of exclusion and vulnerability, as ways of signifying inequalities of status between those who belong and those who don't" (Laborde, 2017, p. 140). According to Laborde, the display of a religious symbol with exclusionary valence by the state, such as the crucifix in schools, violates the third condition of liberal legitimacy that constitutes minimal secularism, namely civic equality.

Scholars are divided on the ECHR's final judgment in *Lautsi*; defenders of the ruling have appealed to reasons including the freedom of the majority to shape its country's policies (Witte, 2011), *realpolitik*'s need to avoid the rise of populism (McGoldrick, 2011), and liberal nationalists' concerns for the protection of a society's historically given national identity (Miller, 2019). I instead intend to ground my defence of the ECHR's judgment in *Lautsi* on Italy's right to self-determination or – in Walzer's words – "communal integrity". The central idea is that any political community has the right to shape its own political institutions and every citizen of that community has the right to live under institutions so shaped (Walzer, 1980, p. 220). Hence, Italians have the right – both as individuals and as a group – to live in a society of an *Italian* sort. The ideal of communal integrity stems from the rights of individuals "to live as members of a historic community and to express their inherited culture through political forms worked out among themselves" (Walzer, 1980, p. 211). Walzer's argument for communal integrity amounts to a defence of the political life, which is unique for each community and "depends upon shared history, communal sentiment, accepted conventions" (Walzer, 1980, p. 228). Italy thus enjoys the presumptive right to have its internal life respected; this includes its tradition to display the crucifix in schools. Surely this communal right has to be balanced against individuals' rights to religious freedom and non-discrimination; yet, as considered in the concurring opinion of

Judge Rozakis and Judge Vajić, the Italian educational context demonstrates toleration towards atheists and members of religious minorities and accords sufficient protection to their rights. Indeed, the wearing of religious symbols by pupils is permitted; whenever possible, schools try to fit in non-majority religious practices; the school curriculum does not include compulsory teaching about Christianity; and optional religious education could be taught to all recognised religious creeds (*Lautsi v. Italy, 2011*: para 74).

It might nonetheless be argued that religious minorities' equal inclusion in the imagined community of citizens ought to be given greater importance. If it is indeed true that symbolic religious establishment may result in citizens' alienation, then the ECHR ought to take this factor into account when balancing the competing interests at stake. Laegaard suggests that symbolic religious establishment can affect civic equality either subjectively or objectively (Laegaard, 2017). Understood subjectively, the effects of symbolic religious establishment are measured in terms of the experiences of minority religions' members. More empirical works needs to be done in this area; yet one recent study fails to find evidence in support of the claim that religious establishment marginalises or alienates religious minorities, thus creating second-class citizens (Perez, Fox and McClure, 2017). Specifically, the authors find no correlation between religious establishment and lower levels of confidence in state institutions.

Understood objectively, on the other hand, symbolic religious establishment is detrimental to civic equality if it is *reasonable* – from a normative standpoint – to regard it as sending a message of exclusion (Laborde and Laegaard, 2019). Indeed, Laborde specifies a person-independent criterion – that is, the impermissibility of a state-endorsed religious symbol does not depend on individuals' subjective feelings of exclusion or whether they "positively associate with the group or identity that is excluded from state endorsement" (Laborde, 2017, p. 140); for the absence of a subjective feeling of alienation is not sufficient as a criterion for establishing the legitimacy of symbolic religious establishment.

Here I disagree with Laborde's view and claim that the effects of symbolic religious establishment ought to be understood subjectively, according to the actual impact that they have on citizens, rather than rejecting them by appeal to some abstract principle (Miller, 2019). I do agree with Laborde that "it is *only* when religious divisions map onto socially salient markers of vulnerability and domination that expressive symbolism" [emphasis added] may be a problem (Laborde, 2017, p. 140 n. 66). Arguably, however, inclusiveness and civic equality are not undermined in a state where all citizens are able to express their religious beliefs in public and to have their rights and freedom recognised. As Miller points out, it is not sufficient to emphasise that religious identity may be a source of social vulnerability in order to conclude that symbolic religious establishment ought to be abolished. It is also necessary to show whether the presence of institutional religious symbols does in fact cause or magnify social vulnerability: to date, this claim is not supported by the evidence (Miller, 2019).

In sum, I defend the ECHR's judgment in *Lautsi* on the grounds that the display of the crucifix in schools may be regarded as a tradition that ought to be respected in virtue of Italy's right to self-determination. The display of the crucifix does not impinge on human rights and there is not enough evidence to conclude that it marginalises or excludes religious minorities from the imagined community of citizens. Hence, I defend the ECHR's decision to grant Italy a wide 'margin of appreciation' as it was not within the its competence to decide on this question. I believe that the value of the ECHR's doctrine of the 'margin of appreciation' lies precisely – to use Walzer's words – in "the respect [that] we are prepared to accord and the room we are prepared to yield to the political process itself. [. . .] It has to do with the range of outcomes we are prepared to tolerate, to accept as presumptively legitimate, though not necessarily to endorse" (Walzer, 1980, p. 229). This is one of the core tenets that my notion of tolerant pluralism seeks to encapsulate.

Conclusion

In this chapter, I have endorsed a revised version of Laborde's minimal secularism – that I have termed 'tolerant pluralism' – and applied it to the headscarf *affaires* and the crucifix case to assess the most debated of ECHR's rulings concerning the appropriate place of religious symbols in the public sphere. I have showed that the headscarf ban in *Sahin* and *Dahlab* is illegitimate because it violates personal liberty as it is justified by a comprehensive view. The prohibition to wear the Muslim veil in *Dogru* also violates personal liberty but on the distinctive grounds that the prohibition infringes on the applicant's ability to live with integrity and is disproportionate to the aims pursued by the law. Finally, I have defended the ECHR's judgment in Lautsi on the grounds that the display of the crucifix in state schools does not result in human rights' violations and is protected by Italy's right to self-determination. In my analysis, I have emphasised some points of disagreements with Laborde's view, for it is my contention that her approach is too secularist with regard to state officials' right to wear – and states' ability to display – religious symbols in the public sphere.

Notes

1 I am grateful to Cécile Laborde, Valentina Gentile and Richard Whatmore for their extensive and precious comments on this paper.
2 I borrow the distinction between personal and institutional symbols from García Oliva and Cranmer, 2013, p. 559. The former reflect an active and voluntary choice of the individual, whereas the latter are somehow imposed on citizens by the state.
3 For Laborde's argument that both separation and establishment can theoretically realise the principles of justice, see Laborde, 2013 (a).
4 Salient freedoms, unlike ordinary ones, relate to "individuals' basic moral powers and their ability to make "strong evaluations" about how to lead their lives" (Laborde, 2017, p. 147). In this case, restrictions of free movement on Shabbat does not limit a salient freedom as it is not connected to an integrity-related interest. But ordinary freedoms can become salient (Laborde, 2017, p. 149).

5 This case is identical to that of *Kervanci v. France*, 2008. For simplicity, I only refer to *Dogru*.

6 The notion of militant democracy was originally formulated by Loewenstein (1937) who argued that democracies had to become militant and enable the government to use the emergency power of restricting fundamental rights for the purpose of protecting the regime from subversive, de-stabilising movements.

7 For a different, yet related discussion of the securitisation of religion and its effects on religious toleration, see Laegaard's contribution to this volume.

8 Here it is worth quoting a passage from *The Idea of Public Reason Revisited*, where Rawls distinguishes between public and secular reasons: "I define secular reasons as reasons in terms of comprehensive nonreligious doctrines. Such doctrines and values are much too broad to serve the purposes of public reason. Political values are not moral doctrines [. . .]. Moral doctrines are on a level with religion and first philosophy. By contrast, liberal principles and values, although intrinsically moral values, are specified by liberal conceptions of justice and fall under the category of the political" (Rawls, 1997, pp. 775, 776).

9 Here Laborde disagrees with me and arrives at a different all-things-considered judgment about the value of liberty, thus concluding that because of her role as teacher of young pupils, the applicant does not have her rights violated.

10 In 2009, the Chamber ruled in favour of Lautsi, thus finding a violation of the afore mentioned Articles.

11 I draw this distinction from Laborde and Laegaard (2019).

References

Bhargava, R. (1998). *Secularism and its critics*. Oxford: Oxford University Press.

Council of Europe (1950). European convention for the protection of human rights and fundamental freedoms (as amended by Protocols Nos. 11 and 14, 4 November 1950). ETS 5. Available at: https://www.refworld.org/docid/3ae6b3b04.html. [Accessed 20 Aug. 2019].

Dahlab v. Switzerland (2001). App. No. 42393/98.

Dogru v France (2008). App. No. 27058/05.

Dworkin, R. (2013). *Religion without god*. Cambridge, MA: Harvard University Press.

Eisgruber, C. and Sager, L.G. (2007). *Religious freedom and the constitution*. Cambridge, MA: Harvard University Press.

Evans, C. and Thomas, C. (2006). Church-state relations in the European court of human rights. *Brigham Young University Law Review*, 3, pp. 699–725.

García Oliva, J. and Cranmer, F. (2013). Education and religious symbols in the United Kingdom, Italy and Spain: Uniformity or subsidiarity? *European Public Law*, 19(3), pp. 555–582.

Laborde, C. (2005). Secular philosophy and Muslim headscarves in schools. *The Journal of Political Philosophy*, 13(3), pp. 305–329.

Laborde, C. (2008). *Critical republicanism: The Hijab controversy and political theory*. Oxford: Oxford University Press.

Laborde, C. (2013a). Political liberalism and religion: On separation and establishment. *The Journal of Political Philosophy*, 21(1), pp. 67–86.

Laborde, C. (2013b). Justificatory secularism. In: G. D'Costa, E. And Malcolm, T. Modood, and J. Rivers, eds., *Religion in a liberal state*. Cambridge: Cambridge University Press, pp. 164–186.

Laborde, C. (2017). *Liberalism's religion*. Cambridge, MA: Harvard University Press.

Laborde, C. and Laegaard, S. (2019). Liberal nationalism and symbolic religious establish-ment. In: G. Guestavsson and D. Miller, eds., *Liberal nationalism and its critics: Empiri-cal and normative dimensions*. Oxford: Oxford University Press.

Lægaard, S. (2017). What's the problem with symbolic religious establishment? The alien-ation and symbolic equality accounts. In: C. Laborde and A. Bardon, eds., *Religion in liberal political philosophy*. Oxford: Oxford University Press, pp. 118–131.

Lautsi v Italy. (2011). App. No. 30814/06.

Leyla Sahin v. Turkey. (2005). App. No. 44774/98.

Lewis, T. (2007). What not to wear: Religious rights, the European court, and the margin of appreciation. *International and Comparative Law Quarterly*, 56(2), pp. 395–414.

Loewenstein, K. (1937). Militant democracy and fundamental rights, I. *American Political Science Review*, 31(3), pp. 417–432. DOI:10.2307/1948164

Macklem, P. (2012). Guarding the perimeter: Militant democracy and religious freedom in Europe. *Constellations*, 19(4), pp. 575–590.

Martínez-Torrón, J. (2012a). Freedom of religion in the European convention on human rights under the influence of different European traditions. *Universal Rights in a World of Diversity: The Case of Religious Freedom. Pontifical Academy of Social Sciences*, pp. 329–355.

Martínez-Torrón, J. (2012b). The unprotection of individual religious identity in the Stras-bourg case law. *Oxford Journal of Law and Religion*, 1(2), pp. 363–385.

McGoldrick, D. (2011). Religion in the European public square and in European public life: Crucifixes in the classroom? *Human Rights Law Review*, 11(3), pp. 451–502.

McKinnon, C. (2007). Democracy, equality and toleration. *The Journal of Ethics*, (11), pp. 125–146.

Miller, D. (2019). What's wrong with religious establishment? In: G. Gustavsson and D. Miler, *Liberal nationalism and its critics: Empirical and normative dimensions*. Oxford: Oxford University Press.

Otto-Preminger-Institut v. Austria (1994). App. No. 13470/87.

Perez, N., Fox, J. and McClure, M. (2017). Unequal state support of religion: On resent-ment, equality, and the separation of religion and state. *Politics, Religion and Ideology*, 18(4), pp. 431–448.

Plesner, I.T. (2005). The European court on human rights between fundamentalist and liberal secularism. Paper presented at the seminar *'The Islamic Head Scarf Contro-versy and the Future of Freedom of Religion or Belief'*. Strasbourg, France, 28–30 July.

Quong, J. (2011). *Liberalism without perfection*. Oxford: Oxford University Press.

Rawls, J. (1997). The idea of public reason revisited. *University of Chicago Law Review*, 64(3), pp. 765–807.

Ringelheim, J. (2012). Right, religion and the public sphere: The European court of human rights in search of a theory? In: L. Zucca and C. Ungureanu, eds., *Law, state and religion in the new Europe: Debates and dilemmas*. Cambridge: Cambridge University Press, pp. 283–306.

Schwartzman, M. (2012). What if religion is not special? *University of Chicago Law Review*, 79(4), pp. 1351–1428.

Schwartzman, M. (2017). Religion, equality, and anarchy. In: C. Laborde and A. Bar-don, eds., *Religion in liberal political philosophy*. Oxford: Oxford University Press, pp. 15–30.

Taylor, C. (1998). Modes of secularism. In: R. Bhargava, ed., *Secularism and its critics*. 6th ed. Oxford: Oxford University Press, pp. 31–53.

Taylor, C. and Maclure, J. (2011). *Secularism and freedom of conscience*. Cambridge, MA: Harvard University Press.

Walzer, M. (1980). The moral standing of states: A response to four critics. *Philosophy & Public Affairs*, 9(3), pp. 209–229.

Witte, J. (2011). Lift high the cross? An American perspective on Lautsi v. Italy. *Ecclesiastical Law Journal*, 13(2), pp. 341–343.

7 The limits of toleration towards Syrian refugees in Turkey

From guesthood to *Ansar* spirit

Ayhan Kaya and Ozan Kuyumcuoğlu

Introduction

The chapter examines the ways in which the Turkish government framed its position towards Syrian refugees by revitalising an act of 'tolerance' and 'benevolence', the origins of which go back to the late 19th and early 20th centuries. Through an analysis of the Ottoman political elite's strategies to prevent the Empire from fragmenting in the late 19th and early 20th centuries (respectively, Ottomanism, Islamism, and Turkism) as well as the speeches of today's political elite, the chapter will highlight how contemporary acts of tolerance and benevolence of the Turkish government draw upon the neo-Ottomanist and Islamist aspirations of the Justice and Development Party (JDP). Based on a discourse analysis of the leading political figures, official texts, legal documents, relevant media coverage, as well as historical scholarship concentrating on the late Ottoman Empire, the chapter will analyse the path-dependence of the rhetoric of tolerance and benevolence embraced by the ruling JDP elite in Turkey vis-à-vis Syrian refugees. It will also, however, scrutinise the ways in which Syrian refugees themselves seek to enter into societal, economic, and political processes in Turkey through a discourse of what Michael Herzfeld (1997, 2016) terms *cultural intimacy*, marking the perception of cultural and religious similarities between themselves and members of Turkish society.

Methodologically, the chapter will make use of legal texts (Law on Foreigners and International Protection enacted in 2014, Regulation on Temporary Protection issued in 2014, Visa Regulations issued by Turkey since the year 2000, either in collaboration with the EU or against EU norms), official statements by leading political figures (such as President Recep Tayyip Erdoğan, former Prime Minister Ahmet Davutoğlu, and former Deputy PM Numan Kurtulmuş), as well as secondary sources on Ottoman history, media archives, and relevant statistics. The speeches by political figures and legal texts will be decoded through discourse analysis (Wodak, 2010). In particular, extracts and quotations from speeches, declarations, and media statements of relevant leading political figures will be examined in order to decode the ways in which Syrian refugees have so far been framed by state actors.[1]

Political approaches in the Ottoman Empire
in the 19th century

In the 19th century, the Ottoman political elite were in search for resurgence after serial defeats against the Great Powers of Europe as well as due to the ongoing structural problems such as economic decline and growing centrifugal nationalist aspirations particularly among the non-Muslim subjects. The Ottoman elite appealed to three political approaches, which became dominant in the Empire after the Imperial Edict of Reorganisation (*Gülhane-i Hattı Hümâyûn or Tanzimat Fermanı*) in 1839. These were Ottomanism, Islamism, and Turkism. All three emerged as a modern reaction to the decline of the Ottoman Empire in the 18th and 19th centuries against the growing hegemony of Western powers. While Ottomanism aimed at creating an Ottoman nation (*ittihad-ı anasır*) that would be inclusive of all the religious and national identities, Turkism aimed at uniting the Ottoman Turks and Turkic populations living in the Caucasus and Central Asia (*ittihad-ı Etrak*). On the other hand, Islamism put forward the idea of the unity of all Muslims (*ittihad-ı İslam*). As discussed later, it became the major political ideology during the reign of II. Abdülhamid. It is useful to discuss Ottomanism and, especially, Islamism in more depth in order to trace their long-term echoes in the political approach of the current Turkish government towards Syrian refugees.

Ottomanism was adopted by the Ottoman governments and political elite of the Tanzimat Era (1839–1878). It was based on equality of all the Ottoman subjects of the Empire, regardless of their national or religious origins. The aim of this political approach was creating a common Ottoman identity by establishing modern institutions and a standardised legal framework, which were supposed to create bonds of citizenship among the imperial subjects, from the Balkans to the Arabian Peninsula. After the Imperial Edict of 1839, Ottoman governments modernised the legal system and created a modern bureaucracy and educational institutions. By the turn of the century, the new bureaucratic class, whose members hailed from different national and religious backgrounds, became the main constituent part of the Ottoman state. Nevertheless, the push of Orthodox Balkan Christians for secession from the Empire during the Ottoman-Russian War of 1877–1878 demonstrated the fragility and limits of Ottomanism. With the loss of territories like Egypt and Cyprus, alongside some parts of Balkans, II. Abdülhamid decided to take control of the bureaucracy after ending the policy of Ottomanisation, putting emphasis, instead, on the unity of Muslims rather than the unity of all imperial subjects. As will be discussed later, Abdülhamid tried to invigorate the bond between Muslims and the state while he was also resorting to suppression towards separatist movements organised by Non-Muslims. During the 'Hamidian era', Islamism became indeed the dominant political approach.

The term 'unity of Muslims' (*ittihad-ı İslam*) was first used in 1872 by a famous Ottoman poet, Namik Kemal, and then by Esad Efendi, who was the clerk of the Commercial Court (Karpat, 2001, p. 27; Özcan, 1997, p. 24). Both of them associated Islam with Ottoman patriotism, which was seen as the only instrument for mobilising Muslim elements against Western aggression to prevent the decline of the Empire. While this understanding was in part an outcome of the social transformations of the late 19th century, with the subsequent efforts of II. Abdülhamid, Islam became the main component of state ideology. Abdülhamid believed that the only way to prevent a possible disintegration of the Empire was to strengthen the unity of all Muslims. Consequently, *Islamism* which refers to the unity of all Muslims (*İttihad-ı İslam not Panislamism*) (Özcan, 1997, p. 24) emerged as a dominant political approach of the Ottoman ruling elite (Karpat, 2001, pp. 17–18). Abdülhamid started an economic reform programme, which aimed to create a Muslim middle class to balance the economic power of Christian subjects who had already become well integrated into the emergent world capitalist networks. There were also ideological consequences to the emergence of Muslim middle classes. Social and economic transformation constrained the mind-sets of the Muslim bureaucrats and intellectuals, who started to search for new methods of social mobilisation based upon Islamic identity (Karpat, 2001, p. 209).

The migration and settlement policies of the Ottoman governments also contributed to the evolution of Muslim middle classes in accordance with the rise of Islamism (Karpat, 2001, pp. 176–177). The Empire had already been a centre of attraction for Muslims living outside of the Ottoman territories before Abdülhamid came to power. Thousands of Algerians escaping from French colonisation were, for instance, welcomed by the Ottoman authorities between 1847 and 1910. Most of them were settled in Syrian cities such as Aleppo, Damascus, and Beirut (Bozpınar, 1997, p. 95), with even some prominent members of the Algerian migrant community appointed to local administrative positions after being granted Ottoman citizenship (Bozpınar, 1997, pp. 105–106; Karpat, 2001, p. 249).

During the first two years of Hamidian rule, hundreds of thousands of Muslims migrated to Ottoman lands, particularly to Anatolia and Northern Syria (Karpat, 2001, p. 339), also due to the disastrous results of the Ottoman-Russian War of 1877–1878, which negatively affected the Empire as well as the Muslims living in the regions occupied by the Russian army. However, this tragedy had also some "constructive side effects" in terms of creating solidarity among Muslims and in beginning to constitute an Islamic political society (Karpat, 2001, p. 229). The popular newspapers of the time, like *Basiret, İttihad, Vakit,* and *Tercüman-ı Hakikat* welcomed the refugees by using a markedly Islamic discourse, echoing in many ways the policies of the current Turkish government towards Syrian refugees. All these newspapers emphasised the importance of the unity of all Muslims and regarded the refugees as *"the guests of Allah (muhajirun)"*. According to the editorial of the *Basiret* newspaper, "unfortunate refugees, whose lands and homes were invaded and destroyed, did not have any other choice then sheltering to the House of the Caliph" (Karpat, 2001, pp. 222–224). Abdülhamid accepted many refugees from various Islamic societies escaping colonial rule with the aim of changing the social structure of the Empire in favour of Muslims; indeed, he also went as far as encouraging Muslims in neighbouring countries to seek refuge on Ottoman soil (Karpat, 2001, p. 341). His openness in accepting refugees was also related to marking the wider, global importance of the Caliphate for the protection and well-being of the world's Muslims (Karpat, 2001, p. 329).

Islamism continued to be an effective political ideology during the Constitutional period despite the efforts of Unionists to bring back Ottomanism. A well-known poet, Mehmed Akif [later Ersoy] and his companions who published the journal of *Sırat-ı Müstakim* [later *Sebilürreşad*] insisted that the ongoing decay of the Empire could only be prevented by strengthening the ties among Muslims (Ersoy, 2011a, p. 12). Despite being a fierce opponent of II. Abdülhamid, Mehmed Akif viewed the Ottoman Empire as the centre of the Islamic world and believed in the unity of all Muslims (Ersoy, 2011b, p. 209). According to him, "*national unity*" could only mean the "indivisible integrity" of Muslims: "You are not Arab, Turk, Albanian, Kurd, Laz nor Circassian. You are only the members of one nation, the great nation of Islam" (Ersoy, 2011a, p. 361). Even after World War I, he continued to support the idea of the "*unity of Muslims*" despite the fact that the "*Arab revolt*" between 1916 and 1918 contributed to the dissolution of the Ottoman Empire (Ersoy, 2011a, p. 56).

Islamism and historical relations with Syria

During the Hamidian rule, Syria, also known as Bilad-Al-Sham (encompassing contemporary Syria, Lebanon, Jordan, Israel and Palestine), took on an important role in reproducing the state ideology. Abdülhamid's state propaganda emphasising the particular status of Arabs in the Empire was so effective that even today the Arab population praises him as one of the greatest political leaders of the past (Karpat, 2001, p. 281). He specifically hired Syrian Arabs as advisors, appointing them to key bureaucratic positions, "*like ministers or secretaries*" (Özcan, 1997, p. 48), and built close relationships with important landowner families such as *Al-Azm* to underline the importance of Syria and to claim Syrians' support against some nationalist organisations and the attempts at infiltration by Western powers in the Arabian Peninsula. In return, despite being inspired by nationalist sentiments, the Syrian political elite viewed the Ottoman state as a shelter against Western colonialist ambitions. Therefore, they decided to comply with the Hamidian regime, which provided them with relative autonomy in political, economic, and social matters (Karpat, 2001, pp. 356–357). Indeed, the Syrian elites of the time were never inclined towards separatist nationalism as long as the Ottoman government tolerated their culture and identity and protected them from foreign aggression. In the eyes of Arabs, the Caliph was the head of Muslims regardless of his national origin (Karpat, 2001, p. 594).

After the Constitutional Revolution of 1908, Syria continued to play an important role in maintaining the integrity of the state and the loyalty of Arabs to the Caliph. Unionists, especially during the Balkan Wars, adopted the propaganda techniques of Abdülhamid to convince the Syrian Arabs to stay within the Empire. However, during the first years of the Constitutional Period, it was only Mehmed Akif that continued to struggle for the unity of Muslims; Unionists, meanwhile, were trying to deal with non-Muslim separatist movements. Mehmed Akif and Islamist intellectuals indeed persistently called for fraternity of Turks and Arabs throughout the Constitutional Period. However, Unionists became convinced to invest more in Turkish-Arab collaboration only after the loss of the Rumelia Provinces during the Balkan Wars of 1912–1913, when Arab nationalism resurfaced again (Khoury, 1983, p. 72). In 1913, Babanzâde İsmail Hakkı, who was an Iraqi Kurd and a well-known and respected Unionist in the Arabian Provinces (Hut, 2013, pp. 106–113), took charge of editorial responsibilities of *Tanin*, the mouthpiece publication of the Committee of Union and Progress. He was chosen by the leaders of the Committee to send a clear message to the Arabs that the Ottoman government was willing to make the necessary political and social reforms which Arabs had been demanding for decades. Accordingly, Babanzâde İsmail Hakkı clearly underlined in his articles that there was no distinction between Turks and Arabs: "Turks always aspire to the happiness of Arabs. Because prosperity of Arabs means the well-being of Turks. Turks and Arabs are destined to protect each other" (İsmail Hakkı, 1913a). He also suggested that the Turkish subjects should be obliged to learn Arabic and highlighted the indivisible integrity of Anatolia and Arab territories (İsmail Hakkı, 1913b). Another prominent Unionist, Halil Hâlid, argued that Turks and Arabs could not live without each other. To him, strengthening the ties between the two elements was essential for preserving the Ottoman lands (and even Islam) from Western aggressors: "the decline of the independence and sovereignty of Islam will be inevitable in case of dissolution of political ties between Turks and Arabs. On such an occasion both will be subject to Western domination and eventually they will lose not only their religion but also their national identities" (Halil Hâlid, 2005, p. 67). The most striking name amongst the Unionists who supported solidarity between Turks and Arabs was Ziya Gökalp, the ideologue of Turkish nationalism. He characterised the Ottoman Empire as a *'Turco-Arab'* state in his famous book *Turkification, Islamization and Civilianization*, which he wrote right after the end of the Balkan Wars in 1913: "the Ottoman state is an Islamic state based on Muslim nations. Two great elements, Turk and Arab nations are the main constituents with their culture and existence. In that sense, the Ottoman State can be characterized as a Turco-Arab state" (Gökalp, 2013, p. 67).

The remarkable point here is the proximity of the Unionists' arguments to those of Mehmed Akif, who had brought forward the idea of resolving the "Arab question" by taking the necessary precautions to ensure the unity of Muslims. This political approach was not only related to a pragmatic approach to resolve the political crisis from which the Empire was suffering. It was also the result of the social transformation, which had started in the middle of the 19th century and accelerated during the Hamidian regime. The growing social mobilisation of Muslims generated new concepts like the unity of Muslims and Islamic brotherhood, which started to dominate the public sphere. It was, of course, not only Ottoman Turks but also Arab intellectuals who became inspired by the political atmosphere prevailing over the Ottoman public opinion. For example, during the disastrous phases of the Balkan War, one Arab journalist called for solidarity between Turks and Arabs, "who must be aware of the danger of Western attempts to split the Empire". [2]

The Islamist discourse of the Unionists continued during World War I. Indeed, they urged Sultan Mehmed Reşad to declare jihad against the Allied forces to galvanise Muslim populations living under the colonial rule of France and Britain (Hanioğlu, 2008, pp. 175–177). While the declaration of jihad did not have a significant effect on Muslims living outside of the Empire, it did succeed in consolidating the loyalty of Syrian Arabs to the Sublime Porte, the Istanbul government. In October 1915, a committee composed of Syrian bureaucrats, journalists, and religious scholars travelled to Anatolian cities and to Istanbul. They also visited the Strait of the Dardanelles, where one of the most iconic battles between the Ottoman army and the Allied powers took place.

Throughout their visit, they met with Ottoman military officers, governors as well as ministers, including the Grand Vizier. Both the members of the Syrian committee and Ottoman officials laid emphasis on the unity of Muslims and fraternity between Turks and Arabs (el-Bâkır et al., 2017, pp. 48–49). Syrian journalist Hüseyin el-Habbâl, from the newspaper *Ebâbîl*, gave a long speech on the loyalty of Syrians to the Caliph and the Ottoman state during the meeting between the Committee members and Interior Minister Talat Pasha. He indicated that "The people of Syria and Palestine are aware of their sacred duty to protect the very existence of the Empire and to remain loyal to the authority of the Caliph over the Holy Lands" (el-Bâkır et al., 2017, p. 73). Habib el-Ubeydî also underlined the necessity of preserving the Islamic ties between Turks and Arabs by pointing out relevant verses in the Quran commanding brotherhood and equality among Muslims (el-Bâkır et al., 2017, p. 123). Some of the committee members even preached in Friday sermons in different mosques in Istanbul (el-Bâkır et al., 2017, p. 127).

A few days later, they visited the Dardanelles Front "to deliver the greetings of the Syrian and Palestinian people to the Ottoman soldiers fighting against the Allied forces" (el-Bâkır et al., 2017, p. 137). Muhammed Kurd Ali, the most well-known Syrian journalist narrated his short-term experience on the Dardanelles front in his article published in *el-Muktebes*. He also described his impressions of Istanbul as a pre-eminent capital due to its location bridging East and West. Kurd Ali also mentioned the beautiful landscape of the city and its cultural characteristics reflecting a synthesis of Eastern and Western civilisations (el-Bâkır et al., 2017, p. 221). His opinions about the "*City of the Caliph (Dârülhilâfe)*" closely reflected the general perceptions of Syrian Arabs' of Istanbul; perceptions still relevant today.

This historical overview of the social and political transformations of the 19th and early 20th centuries is important in highlighting the long history of close relations between Ottoman Turkey and Syria, and thus crucial to understanding also the contemporary appeals to that very past. As we noted, the transformations of the 19th century were key to generating a political consciousness of Muslim identity, which became not only a component of political Islam but also a part of Turkish and Arab nationalism. That is why, despite the subsequent emergence of state nationalism in both Anatolia and Syria, cultural and traditional connections between the two populations persisted. Attempts to form a Turco-Arab Empire after the Balkan Wars (Kayalı, 2003, p. 151) and collaboration between Turk-ish and Arab nationalists in Southern Anatolia and Northern Syria against the French forces from 1919 to 1921 (Cebesoy, 2010, pp. 354–357) also foreground that a strong traditional understanding of Muslim solidarity emerged out of both the social transformations of the 19th century, and the efforts of Abdülhamid to create a Muslim nation. Despite the turn towards nationalist state building pro-jects in both Turkey and Syria in the 1920s, a close relationship between the two countries continued in many respects. For instance, after the proclamation of the Turkish Republic in 1923, Turkey continued to support independence movements in Syria. In fact, it accepted many Syrian refugees escaping both from the French-occupying forces during the "Great Revolt" of 1925 as well as from the small-scale civil war that took place among some ethnic groups in Northern Syria in 1937 (Umar, 2003, pp. 196–198).

The vocabularies of migration politics in contemporary Turkey

Following the longer historical overview, we will now turn to a discussion of more recent histories of relations between the Turkish state and migrant popula-tions from neighbouring lands, Muslim and not. A first official Law on Settlement was adopted in response to the arrival of ethnic Turks already in the early years of Republic.[3] It has continued to be the main legislative text dealing with immigra-tion, and it determines who can enter, settle and/or apply for refugee status in Tur-key. However, it also provides individuals of Turkish descent and culture with the opportunity to be accepted as 'migrants' and refugees in Turkey (İçduygu 2015). For instance, Uzbeks, Turkomans, Bulgarian-Muslims, and Uighurs migrating to Turkey from different parts of the world are labelled as 'migrants' (*göçmen* in Turkish) in the official documents as well as in everyday life, as they are ethni-cally of Turkish descent.

Alongside this distinction, there are two other terms pertaining to the categories through which migration to Turkey is framed which require further elaboration: the terms "guest" (*misafir*) and "foreigner" (*yabancı*). In the official literature, the term "guest" has been hitherto used to refer to refugees with Muslim origin but without Turkish ethnic origin, coming from outside the European continent. Kurdish refugees in the 2000s and Syrian refugees in the 2010s were referred to as 'guests' since Turkey officially does not accept refugees coming from outside its western boundaries. Bosniac and Kosovar refugees seeking refuge in Turkey in the 1990s constituted an exception as they were coming from beyond the west-ern borders of Turkey, and had the right to apply for asylum in Turkey accord-ing to the geographical limitation clause that Turkey decided to maintain[4] in the 1967 Additional Protocol of the Geneva Convention on the protection of refugees (that removed geographical limitations). On the other hand, the term 'foreigner' is most often used in official texts as well as in public discourse to refer to those who are neither Turkish nor Muslim. These groups cannot also be incorporated into the prescribed national identity, which is mainly based on a 'holy trinity' of

Sunni-Muslim-Turkish elements. Accordingly, not only the non-Muslims coming from abroad but also autochthonous groups such as Greeks and Armenians resident in Turkey that are named as "foreigners", or "local foreigners" in legal texts (Çetin, 2002).

Traditionally known as emigration countries, over the past few decades, Turkey, Lebanon and Jordan have also become key settlement and transit spaces for economic and forced migrants (De Bel-Air, 2006; Pérouse, 2013). Syrian refugees in particular have been considered 'guests' not just by the Turkish but also by the Lebanese and Jordanian states. From the very beginning of the refugee arrivals, the 'welcoming' of Syrians was presented by the host states and societies as based on deep-rooted values such 'Turkish hospitality', 'Muslim fraternity', 'Arab hospitality', and 'guesthood' traditions (De Bel-Air, 2006; Pérouse, 2013; Chatty, 2013; El Abed, 2014; Kirişçi, 2014; Erdoğan, 2015; Erdoğan and Kaya, 2015; Baban, Ilcan and Rygiel, 2016). These deep-rooted values were tied to common historical experiences such as the refugee flows from Algeria, the Caucasus, and the Balkans towards the Ottoman lands in the 19th century, described earlier. As we noted previously, indeed, Muslim refugees had been welcomed by the Ottoman ruling elite and society as the '*guests of Allah*'. They were not only settled in Anatolia but also in the Bilad-Al-Sham region, encompassing contemporary Syria, Lebanon, Palestine, Israel, and Jordan. These entangled histories are one of the reasons why the Turkish, Jordanian, and Lebanese governments appeal to 'historical and cultural ties' with the Syrian refugees rather than framing the issue of reception through international conventions.

Nevertheless, as powerful as such historical associations may still be, they also mark the temporary and precarious nature of the willingness of the state to accept refugees or guests. In this sense, a more recent metaphor adopted to qualify the role that the Turkish state and pious Muslim-Turks should play for Syrians in Turkey has been that of the *Ansar* spirit (Arabic for 'helpers'). As a metaphor, *Ansar* refers to the people of Medina who supported the Prophet Mohammad and the accompanying Muslims (*muhajirun*, or migrants) who migrated there from Mecca, which was under the control of pagans. The metaphor of *Ansar* originally points at a temporary situation as the Muslims later returned to Mecca after their forces recaptured the city from the pagans (Korkut, 2015).[5] Hence, the Turkish government has appealed directly to Islamic symbolism to legitimise its acts in responding the Syrian refugee crisis. Government leaders have, indeed, consistently compared Turkey's role in assisting Syrian refugees to that of the *Ansar*, referring to the Medinans who helped Muhammad and his entourage. Framing the reception of Syrian refugees within the discourse of *Ansar* and *Muhajirun* has thus served to elevate public and private efforts to accommodate Syrian refugees from a humanitarian responsibility to a religious and charity-based duty (Erdemir, 2016).

Then the PM Ahmet Davutoğlu, in his speech in Gaziantep, one of the most popular destinations for Syrian refugees on the Syrian-Turkish border, publicly stated that the inhabitants of Gaziantep are a city of *Ansar*:

Gazi[antep] is an Ansar city now. God, bless you all.[6]

Similarly, President Erdoğan used the same discourse in his speeches in 2014 and after:

> In our culture, in our civilization, guest means honour, and blessing. You [Syrian guests] have granted us the honour of being Ansar, but also brought us joy and blessing. As for today, we have more than 1,5 million Syrian and Iraqi guests.[7]

The discourse of Ansar has continued until recently, with Deputy PM Numan Kurtulmuş, referring to the very same rhetoric when he introduced the right to work granted to Syrian refugees under temporary protection:

> The reason why the Syrian refugees are now settled in our country is the hospitality and Ansar spirit that our nation has so far adhered to. Other countries cannot do anything when encountered with a few hundred thousands of refugees. But contrary to what the rich and prosperous countries could not do for the refugees, our country did its best for the refugees as a generous host, friend, brother and neighbour.[8]

What can be noted, then, is that the main common denominator of the ruling political elite is that Syrian refugees are being portrayed and framed by means of an act of benevolence. The assistance of the state to refugees is therefore based on charity, rather than any universally recognised rights that are supposed to be granted to refugees fleeing their homelands. Indeed, as was noted earlier, the Turkish state does not name Syrian refugees as 'refugees'. State actors thus engage the issue not through universal law but through the laws of religious charity and benevolence. It is striking, indeed, that also bureaucrats working in the migration sector have embraced such a religious-based discourse concerning the reception of Syrian refugees in Turkey.

At the same time, it is important to note that the Turkish government embraced the refugees as "*guests*" not only because of religious aspirations but also to reproduce its political hegemony by utilising the notion of the "unity of Muslims". As we argued previously, the same policy was followed by II. Abdülhamid in the late 19th and early 20th centuries to transform the social structure of the Empire in favour of Muslim subjects, and to underline the central role of the Caliph to the rest of the world's Muslims. Also, this similarity points to important continuities in Islamic rhetoric related to Muslim refugees from the Ottoman past to contemporary Turkish political life. It also presents a perspective for understanding the background of the contemporary Turkish government's attempts to revitalise the idea of "Islamic brotherhood" by drawing upon the "glorious era of II. Abdülhamid" and Mehmed Akif Ersoy's speeches and poems[9] (Sabah, 2018a).

Syrian-Turkish relations and the discourse of cultural intimacy

In studies carried out by the authors and other scholars, Syrian refugees have reported that they are relatively content with their residence in Turkey. Indeed, such studies reveal that ethno-cultural, religious, and historical ties between most of the Syrians and the native Turkish citizens have been reported by Syrian inter-locutors to be the main source of 'comfort' for their stay in Turkey. This is what Michael Herzfeld (2016) refers to as "cultural intimacy", which serves as a form of reassurance for Syrian refugees to remain in Turkey despite social-economic difficulties, deprivation of rights, exclusion, and exploitation in the labour market as well as in everyday life.

It seems that it is partly such "cultural intimacy" that prevents most of the Syrians from generating a willingness to go to Europe. An extensive study conducted by Kaya and Kıraç (2016) in Istanbul revealed that only 1.6% of the interviewed Syrians were willing to go to Europe, while 79% expressed their willingness to go back to Syria, and around 20% stated their willingness to stay in Turkey when the war is over. Similarly, Fabbe, Hazlett, and Sinmazdemir (2017) have also revealed a similar tendency among the Syrian refugees surveyed in Gaziantep, Urfa, Hatay, and Istanbul. In their survey, they found that only around 5% of the Syrian refugees were willing to go to Europe. Their hesitation to go to Europe can be explained through various factors: cultural intimacy with Turkey, ethnic and religious affinity with the natives in Turkey, most of the Syrians being Sunni-Muslim-Arab who have communal, religious, and ethnic ties in Turkey (especially in the southeast of Turkey as well as in Istanbul), growing anti-refugee sentiments in Europe, absence of safe passage to Europe, the obvious risks involved in the sea crossing, the economic burden of the journey, and the news with regards to the deadly journeys circulated in both social and mainstream media. These figures are striking also for how they contrast with prevailing popular understandings among European publics that are, for the great part, utterly unaware of Syrian refugees' orientations towards a potential journey to, and permanence in, Europe.

In his path-breaking ethnographic book, *Cultural Intimacy: Social Poetics in the Nation-State*, Michael Herzfeld (1997; Second Edition 2005) elaborated on what he called forms of "cultural intimacy" in Greece, marking a strong sense of difference between what they presented to the 'outside world' and what they knew about them-selves on the 'inside'. Herzfeld defines cultural intimacy as "the recognition of those aspects of a cultural identity that are considered a source of external embarrassment but that nevertheless provide insiders with their assurance of common sociality" (Herzfeld, 2005, p. 3). He subsequently expands this notion by drawing our attention to the fact that the term "cultural intimacy" was often perceived in the literature as simply the idea of acquaintance and familiarity with a culture (Herzfeld, 2013, p. 91).

Adopting this Herzfeld's conception, we can note that Arabic-speaking Sunni Syrians have already created 'comfort zones' in various districts of Istanbul, which seem to provide them with a cultural intimacy with regards to religious, moral, architectural, urban, and sometimes even linguistic similarities originating from the common Ottoman past of the Turks and the Arabs. In that regard, the speeches and articles of prominent Syrian political elites who visited the cities of Anatolia and Istanbul in 1915 make for important historical examples. As discussed pre-viously, such narrators very frequently highlighted "the symbolic importance" of Istanbul, "*the City of the Caliph (Dârülhilâfe)*" and explicitly expressed their pride in being part of an "Islamic fraternity under the holy crescent".

As suggested earlier, Herzfeld's notion of cultural intimacy does not only refer to "the sharing of known and recognizable traits" with the ones inside, but it also refers to those traits "disapproved by powerful outsiders" (Herzfeld, 2016; Byrne, 2011, p. 148). It could be argued that this second component of cultural intimacy comes into play when the Syrian refugees residing in Istanbul as well as in other parts of Turkey, especially in the southeastern parts of the country, are asked to express their opinion about whether they want to migrate to Europe. Syrian refugees interviewed in both of the above cites studies predominantly expressed their willingness to stay in Istanbul, Gaziantep, Urfa, and Hatay, and not to go to Europe (Kaya and Kıraç, 2016; Fabbe, Hazlett and Sinmazdemir, 2017). Their hesitation to go to Europe seems to derive partly from their strong belief that they are considered with disapproval by Europeans, and partly from the life-threatening nature of the journey, which has already led to the death of thousands of people *en route*. It was often put forward by the interlocutors during the field research that the tragedies that their fellow Syrians had to endure during their exodus from Syria to Greece have left very negative marks on them. The traces of the heart-breaking images of Ailan Kurdi, whose dead body was laying down on the Aegean shores of Bodrum, Turkey,[10] were still fresh in the minds of the interlocutors when interviewed. At the same time, many Syrians' hesitation to go to Europe may also be explained by their deeply rooted suspicion towards Europeans coming from historical experiences under the French mandate regime in Syria between 1919 and 1936. During the mandate era, as the French administration decided to ignore the basic demands of the local population and tried to suppress all kinds of opposition movements, Syrians rejected to comply with the mandate order and many of them decided to seek refuge in Turkey, especially during the 'Great Revolt' of 1925.

Herzfeld's notion of cultural intimacy includes various acts and attitudes repeated by members of a group of people, which lead to the formation of a Manichean understanding of the world divided between 'us' and 'them'. These acts and attitudes may range from essentialising a culture and a past, practising various stereotypes in everyday life, performing persuasive acts of resemblances, ordinary acts of embarrassment kept as intimate secrets of the group, and different forms of *iconicity*[11] such as mythical, visual, musical, and gastronomic images bridging a sense of resemblance with the other members of the group at large (Herzfeld, 2016). According to Herzfeld, the essentialisation and reification of a past and culture do not only serve as an ideological element instrumentalised by political institutions and states to control and manipulate the masses, but are also an indispensable element of social life (Herzfeld, 2016, p. 33). Hence, ordinary individuals also tend to essentialise and reify the past for their own use to come to terms with the hardships of everyday life.

Essentialising the past partly makes it possible for private individuals to create the semiotic effect of what Herzfeld calls *iconicity*, the principle of signification by resemblance, which contributes to the formation of the Manichean understanding of a world divided between 'us' and 'them' noted earlier (Herzfeld, 2016, p. 33). Mythical, visual, musical, heroic, even gastronomic iconicities are all likely to contribute to the formation of this Manichean worldview. Syrian refugees residing in Istanbul at large, and in conservative neighbourhoods especially, are likely to construct bridges between themselves and the members of the majority society by means of visual, musical, religious, gastronomic, and even linguistic iconicities, which create a space of intimacy with the host communities. Historical processes and experiences such as the emergence of a Muslim middle class accompanied by a sense of Muslim solidarity in the Ottoman Empire of the late 19th century, undoubtedly contributed in the creation of such iconicities.

Nevertheless, the framing of the refugee reality by Turkish state actors as an act of 'benevolence' and 'tolerance' has also shaped public opinion in a way that has led to the exposure of some racist and xenophobic attitudes vis-a-vis refugees. Some retrospective controversies highlighted by nationalist discourses play a crucial role in provoking hostility against the Syrians. Historical events like the '*Arab revolt*' initiated by Sharif Hussein against the Ottoman army in 1916 have been instrumentalised by nationalist circles to accuse the Arabs of treason, although a great proportion of the Ottoman Arabs remained loyal to Istanbul until the end of World War I. Such distorted historical narratives seem to have an important impact on some parts of the Turkish population's perception of Syrian refugees. Indeed, there have been several lynching attempts, stereotypes, prejudices, communal conflicts, and other forms of harassments performed against Syrians (Gökay, 2015; Togral-Koca, 2016).

The massive increase in the number of refugees living outside of camps and the lack of adequate assistance policies toward them has also aggravated a range of social problems. Refugees have experienced various problems of adaptation in big cities and the language barrier has seriously complicated their ability to integrate into Turkish society. There are several problems that the Syrians have been facing in everyday life. There is now a growing concern about underage Syrian girls being forced into marriage as well as fears that a recent constitutional court ruling decriminalising religious weddings without civil marriage will lead to a spread of polygamy involving Syrian women and girls (Kirişçi and Ferris, 2015). The sight of Syrians begging in the streets has also caused particular resentment among local people, especially in the western cities of Turkey. There have also been reports of occasional violence between refugees and the local population. This, in turn, has reinforced a growing public perception that Syrian refugees are associated with criminality, violence, and corruption. These attitudes contrast with the observations of local authorities and security officials. In reality, criminality has been surprisingly low, and Syrian community leaders have been very effective in preventing crime and defusing tensions between refugees and locals (Kirişçi and Karaca, 2015).

In times of economic and political instability, nationalist and populist agenda become more visible, and authorities attempt to generalise hostility towards others who are culturally, ethnically and religiously different. Refugees are easily portrayed as inferior, maligned, dangerous, or threatening.[12] Lacking the resources of public communication and relevant language skills, most of the refugees are unable to contest such labellings, stereotypes, and xenophobic attitudes generated by the majority society (Marfleet, 2007, 2013). Such a xenophobic discourse was also employed by the main oppositional parties prior to the June 7, 2015, General Elections in their electoral campaigns. The Republican People's Party (CHP) and the Nationalist Action Party (MHP) used a populist discourse scapegoating the Syrian refugees for the political, social, and economic ills in Turkey. The Syrian refugees were instrumentalised by both parties to express their critique against the JDP, which they blamed for deepening the Syrian crisis in the first place, thus leading to the massive migration of Syrians to Turkey at the expense of Turkish citizens (Werz, Hoffman and Bhaskar, 2015). Upon growing criticisms coming from civil society organisations and academics, it should also be noted here that both parties, especially the CHP, gave up on such discourses prior to the second general elections held on November 1, 2015, and have since then used a rather constructive and friendly discourse vis-a-vis the Syrians (Canyaş, Canyaş and Gümrükçü, 2015). However, the same anti-refugee discourse has been repeated by the CHP leader in response to the government's efforts to grant citizenship to Syrians prior to the constitutional change referendum on April 16, 2017.[13]

Temporary, or permanent 'guests'?

With the passing of time, the framing of Syrian refugees as 'guests' was no longer sustainable, both in terms of accommodating their urgent needs, as well as in coming to terms with the increasing resentment among the local populations vis-à-vis the refugees. Turkey first introduced a Temporary Protection Directive for the refugees in 2014, based on Articles 61 to 95 of the Law on Foreigners and International Protection, which came into force in April 2014. The directive grants Syrians almost the entire spectrum of social and civil rights that refugees enjoy in Western societies.[14] Accordingly, Turkey has provided Syrians with temporary protection, which consists of three elements: an open-door policy for all Syrians, no forced returns to Syria (non-refoulement), and an unlimited duration of stay in Turkey.

Following the implementation of the Temporary Protection regulation, which still frames the refugees with a state of temporariness, some discursive shifts were witnessed in the media regarding state actors' changing position on the permanent character of at least some of the Syrian refugees in Turkey. These discursive shifts have so far mainly emphasised the permanent nature of the issue – the introduction of work permits in early 2016, the incorporation of pupils into public schools, creating quotas for Syrian students in higher education institutions, granting citizenship to Syrians, and some statements from political figures such as President Erdoğan and Deputy PM Numan Kurtulmuş. Comparing Turks living in Germany and the United States with Syrians living in Turkey, in a meeting with journalists President Erdoğan referred to the need for granting citizenship to Syrian refugees residing in Turkey:

> Today, a Turk can go to Germany and become a German citizen; [a Turk] can go to the U.S. and become an American citizen; why can't the same be possible for people living in our country?[15]

This statement brought about a big commotion in Turkey, raising fears that all the Syrians would be granted citizenship immediately. Following public concern, Deputy PM Kurtulmuş had to announce that the Ministry of Interior was working on a proposal in this direction, implicitly signifying that the government considered granting citizenship to those with cultural and economic capital:

> Our citizens should be comfortable. We have not yet completed the proposal about granting citizenship to the Syrians. The Ministry of Interior is working on the proposal. There are so many skilled people [among the Syrians] who can contribute to Turkey. To this effect, we can propose some criteria. When there is nothing concrete, some oppositional groups are trying to create chaos for the sake of opposition; and these groups are gossiping about the uncertain things as if everything is clearly laid out by the government. These are all incorrect.[16]

However, it is not still clear what Turkish state actors intend by granting citizenship. Anecdotal evidence suggests that those Syrians with economic and cultural capital are more likely to be granted citizenship than precarious ones who seem to be instrumentalised by the ongoing neo-liberal forms of governance for the establishment of a model of precarious work for non-citizen workers (Canefe, 2016; Baban, Ilcan and Rygiel, 2016).

How can we place this latest evolution of the framing of the refugee issue? In this chapter, we have attempted to provide a longer-term perspective on the ways in which Turkish state actors have perceived Syrian refugees since their exodus to Turkey in 2011. The argument made in the chapter is that the JDP political leadership has revitalised a neo-Ottomanist and Islamist discourse, the origins of which can be traced back to the late 19th-century Ottoman Empire. Based on an analysis of late Ottoman history, legal texts, and the speeches of leading political actors, we suggest that the JDP leadership has reproduced an Islamist discourse in' their attempts to incorporate Syrians on the basis of the principle of tolerance and benevolence of Turkish state actors, who tend to see their Sunni-Muslim brothers and sisters as the members of the same Millet, the community of faith. These acts of benevolence go in parallel with the discourse of *Ansar* spirit, reminding the leading political elite of the early Muslims of Medina welcoming the Prophet Mohammad and his entourage escaping from the atrocities of their pagan relatives in Mecca. It was argued that it is such framings of benevolence, coupled with perceptions of a shared cultural intimacy, that have served to provide a relatively 'tolerant' context for many Syrian refugees in Turkey. Whether this political discourse of tolerance and cultural intimacy embodied by the JDP government is positively perceived by the overall Turkish society is an open question. Growing societal tensions between local communities and Syrians show that the majority of Turkish society that is exposed to growing socio-economic and political challenges no longer embraces the political discourse of tolerance and cultural intimacy.

Notes

1 The chapter draws on field research conducted by Ayhan Kaya within the framework of the Horizon 2020 research project 'RESPOND: Multilevel Governance of Mass Migration in Europe and Beyond', www.respondmigration.com/. Part of the research was also undertaken within the Horizon 2020 project ISLAM-OPHOB-ISM, Grant Agreement ERC AdG785934.

2 *İkdam* (1913). "Suriye Islâhâtı" (Reform in Syria).

3 *İskan Kanunu* (Settlement Law), Law No. 2510 of 1934, provides that only migrants of Turkish culture, with an objective of settling in Turkey, can obtain immigrant status (Art. 3), and that those of non-Turkish origin will not be accepted as immigrants in Turkey (Art. 4). This Law was amended in 2006 but its main understanding of who can be an immigrant has not been substantially altered. See the reformed Law No. 5543 on Settlement (*İskan Kanunu*) of 26 September 2006 at www.nvi.gov.tr/Files/File/Mevzuat/Nufus_Mevzuati/Kanun/pdf/IskanKanunu.pdf [last visited 17 January 2019]

4 Together with Congo, Madagascar, and Monaco.

5 Haber7 (2014). "Cumhurbaşkanı Erdoğan: Bizler Ensar sizler muhacir" (The President Erdoğan: We are Ansar you are muhacir), www.haber7.com/ic-politika/haber/1208342-cumhurbaskani-erdogan-bizler-ensar-sizler-muhacir accessed on 4 January 2019.

6 *Akşam*, 28 December 2014, www.aksam.com.tr/siyaset/davutoglu-gazi-sehir-artik-ensar-sehirdir/haber-367691 accessed on 7 November 2018.

7 *Hurriyet*, 8 October 2014, www.hurriyet.com.tr/erdogan-suriyeli-siginmacilara-seslendi-27342780 accessed on 7 November 2018.

8 *Ajans Haber*, 11 January 2016, www.ajanshaber.com/bakanlar-kurulu-sonrasi-kurtulmustan-aciklama-haberi/325379, accessed on 8 November 2018.

9 *Sabah* (2018a). "Cumhurbaşkanı Erdoğan: Abdülhamid Han Dünyanın Son Evrensel İmparatorudur", (President Erdogan: Abdülhamid was the last Universal Emperor), www.sabah.com.tr/gundem/2018/02/11/cumhurbaskani-erdogan-abdulhamid-han-dunyanin-son-evrensel-imparatorudur, accessed on 26 February 2019; *Sabah* (2018b). "Cumhurbaşkanı Erdoğan: Mehmet Akif'i Geleceğe Taşımalıyız", (President Erdogan: We must Carry Mehmet Akif into Future), www.sabah.com.tr/gundem/2018/12/26/cumhurbaskani-erdogan-mehmet-akifi-gelecege-tasimaliyiz, accessed on 26 February 2019.

10 H. Smith. 2015. "Shocking images of drowned Syrian boy show tragic plight of refugees," *The Guardian* (2 September), available at www.theguardian.com/world/2015/sep/02/shocking-image-of-drowned-syrian-boy-shows-tragic-plight-of-refugees, accessed on 20 January 2019.

11 The term 'iconicity' derives from the word Greek word *eikôn*, which literally means likeness conveying the imitation of the divine. However, in modern usage, the term 'icon' carries a misleading meaning, which is often interchangeably used with 'emblem'. Michael Herzfeld's use of the term 'iconicity' derives from its original meaning, which has something to do with resemblance. For further explanation on the term, see Herzfeld (2016, p. 93).

12 For a detailed account of the ways in which refugees have been portrayed in such terms see Wodak and van Dijk (2000), Walters (2006), and Inda (2006).

13 For Kemal Kılıçdaroğlu's words on the Syrians, see *Hurriyet Daily News* (March 6, 2017), "If one person makes mistake, the whole country will have to pay for it: CHP leader," available at www.hurriyetdailynews.com/if-one-person-makes-mistake-the-whole-country-will-have-to-pay-for-it-chp-leader.aspx?pageID=238&nID=110283&NewsCatID=338 accessed on 6 March 2019.

14 For the text of the Geneva Convention and Protocol relating to the Status of Refugees, see www.unhcr.org/3b66c2aa10

15 For news coverage about President Erdoğan's discourse on the Syrians being granted citizenship, or dual nationality, see *Hurriyet Daily News* (July 11, 2018), www.hurriyetdailynews.com/erdogan-details-dual-citizenship-for-syrians.aspx?pageID=238&nid=101428&NewsCatID=341.

16 For the Deputy PM Numan Kurtulmuş's speech on granting citizenship to the Syrians see *Sabah* (July 15, 2016), www.sabah.com.tr/gundem/2016/07/14/hukumetten-suriyelilere-vatandaslik-aciklamasi.

References

Baban, F., Ilcan, S. and Rygiel, K. (2016). Syrian refugees in Turkey: Pathways to precarity, differential inclusion, and negotiated citizenship rights. *Journal of Ethnic and Migration Studies*, 43, pp. 41–57.

Bozpınar, Ş.T. (1997). Suriye'de Yaşayan Cezayirli Mültecilerin Tâbiyeti Meselesi (1847–1900) (The Question of Citizenship of the Algerian Immigrants Settled in Syria, 1847–1900). *İslâm Araştırmaları*, 1, pp. 91–106.

Byrne, D. (2011). Archaeological heritage and cultural intimacy: An interview with Michael Herzfeld. *Journal of Social Archaeology*, 11(2), pp. 144–157.

Canefe, N. (2016). Management of irregular migration Syrians in Turkey as paradigm shifters for forced migration studies. *New Perspectives on Turkey*, 54, pp. 9–32.

Canyaş, F.B., Canyaş, O. and Gümrükçü, S.B. (2015). Turkey's 2015 parliamentary elections. *Journal of Balkan and Near Eastern Studies*, 18(1), pp. 77–89.

Cebesoy, A.F. (2010). *Milli Mücadele Hatıraları*, (Memoirs from the Independence War, Osman Selim Kocahanoğlu, ed.). 2nd ed. Istanbul: Temel.

Çetin, F. (2002). Yerli Yabancılar. In: I. Kaboglu, ed., *Ulusal, Ulusalüstü ve Uluslararası Hukukta Azınlık Hakları*. Istanbul: Istanbul Barosu Insan Hakları Merkezi Yayınları, pp. 70–81.

Chatty, D. (2013). Guests and hosts: Arab hospitality underpins a humane approach to asylum policy. *The Cairo Review of Global Affairs*, 19, Apr. 22.

De Bel-Air, F. (2006). *Migration and politics in the Middle East: Migration policies, nation building and international relations*. Beyrouth: IFPO.

El Abed, O. (2014). The discourse of Guesthood: Forced migrants in Jordan. In: A. Fabos and R. Osotalo, eds., *Managing Muslim mobilities*. London: Palgrave.

el-Bâkır, Muhammed, Muhammed Kürd Ali, Hüseyin el-Habbâl, Abdülbâsıt el-Ünsî (2017). *Türklerle Omuz Omuza: Arap İlim Cemiyeti Dârülhilâfe ve Çanakkale'de* (Ali Benli, trans.). İstanbul: Klasik.

Erdemir, A. (2016). The Syrian refugee crisis: Can Turkey be an effective partner? *Foundation for defence of democracies*, 16 Feb. 2016. Available at: http://www.defenddemocracy.org/media-hit/dr-aykan-erdemir-the-syrian-refugee-crisis-can-turkey-be-an-effective-partner/ [Accessed 3 Oct. 2016].

Erdoğan, E. (2014). Unwanted, unwelcome: Anti-immigration attitudes in Turkey. *Working Paper*, German Marshall Fund of the United States, 10 Sept.

Erdoğan, M. (2015). *Türkiye'deki Suriyeliler* (Syrians in Turkey). Istanbul: Istanbul Bilgi University Press.

Erdoğan, M. and Kaya, A., eds. (2015). *14. Yüzyıldan 21. Yüzyıla Türkiye'ye Göçler* (Migration to Turkey since the 14th Century). Istanbul: Istanbul Bilgi University Press.

Ersoy, M.A. (2011a). *Düzyazılar: Makaleler-Tefsirler-Vaazlar*, (Selected Prose of Mehmed Akif Ersoy: Articles, Tafseers and Preaches, A. Vahap Akbaş, ed.). 2nd ed. İstanbul: Beyan.

Ersoy, M.A. (2011b). *Safahat*, (Phases, Mehmet Dikmen, ed.). İstanbul: Bağcılar Belediyesi.

Fabbe, K., Hazlett, C. and Sinmazdemir, T. (2017). What do Syrians want their future to be? A survey on refugees in Turkey. *Foreign Affairs*, 1 May.

Gökalp, Z. (2013). *Türkleşmek, İslamlaşmak, Muasırlaşmak*, (Turkization, Islamization, Civilianization, Yalçın Toker, ed.). 6th ed. Istanbul: Toker.

Gökay, B. (2015). The making of a racist myth. *Ottomania*, 15 Jan. Available at: http://www.fabrikzeitung.ch/the-making-of-a-racist-myth/

Halil Hâlid, Ç. (2005). *Türk ve Arap*, (Turk and Arab, M. Ertuğrul Düzdağ, ed.). 3rd ed. Istanbul: Yeni Zamanlar.

Hanioğlu, M.Ş. (2008). *A brief history of the late Ottoman empire*. Princeton, NJ: Princeton University Press.

Herzfeld, M. (2005). *Cultural intimacy: Social poetics in the nation-state*. 2nd ed. New York: Routledge.

Herzfeld, M. (2013). The European crisis and cultural intimacy. *Studies in Ethnicity and Nationalism*, 13(3), pp. 491–497.

Herzfeld, M. (2016). *Cultural intimacy: Social poetics and the real life of states, societies and institutions*. 3rd ed. New York: Routledge (first edition: 1997).

Hut, D. (2013). II. Meşrutiyet Döneminde Bir Osmanlı Entelektüeli, Babanzade İsmail Hakkı (1876–1913). (An Ottoman Intellectual during the Constitutional Period, Babanzade İsmail Hakkı), *Tarihimizden Portreler: Osmanlı Kimliği, Prof. Dr. Cevdet Küçük Armağanı*, (Portraits from Turkish History: Ottoman Identity, Zekeriya Kurşun, Haydar Çoruh, eds.). Istanbul: ORDAF, pp. 101–122.

İçduygu, A. (2015). Turkey's evolving migration policies: A Mediterranean transit stop at the doors of the EU. *IAI Working Paper* 15/31, Rome, Sept.

İkdam (1913). Suriye Islâhâtı. Jan. 6.

Inda, J.X. (2006). *Targetting immigrants: Government, technology and ethics*. Oxford: Blackwell Publishing.

İsmail Hakkı, B. (1913a). Arabca ve Islâhât (Arabic Language and Reform). *Tanin*, 21 April.

İsmail Hakkı, B. (1913b). Beyrut Hadisatı (Developments in Beirut). *Tanin*, 19 April.

Karpat, K. (2001). *İslâm'ın Siyasallaşması*. (Politization of Islam) Istanbul: Istanbul Bilgi University.

Kaya, A. and Kıraç, A. (2016). Vulnerability assessment of Syrian refugees in Istanbul. Report, support to life association. Available at: http://eu.bilgi.edu.tr/media/files/160621_Ist_NA_Report.pdf

Kayalı, H. (2003). *Jön Türkler ve Araplar: Osmanlıcılık, Erken Arap Milliyetçiliği ve İslamcılık (1908–1918)* (Türkan Yöney, trans.). 2nd ed. Istanbul: Tarih Vakfı.

Khoury, P.S. (1983). *Urban notables and Arab Nationalism*. New York: Cambridge University Press.

Kirişçi, K. (2014). Syrian refugees and Turkey's challenges: Going beyond hospitality. *Working Paper*, May, Brookings Institute, Washington, DC.

Kirişçi, K. and Ferris, E. (2015). Not likely to go home: Syrian refugees and the challenges to Turkey— and the international community. *Working Paper*, No. 7, Sept., Brookings Institute, Washington, DC.

Kirişçi, K. and Karaca, S. (2015). Hoşgörü ve Çelişkiler: 1989, 1991 ve 2011'de Türkiye'ye Yönelen Kitlesel Mülteci Akınları. (Tolerance and Conflicts: Massive Refugee Flows to Turkey). In: M. Erdoğan and A. Kaya, eds., *14. Yüzyıldan 21: Yüzyıla Türkiye'ye Göçler* (Migration to Turkey since the 14th Century). Istanbul: Istanbul Bilgi University Press, pp. 295–314.

Korkut, U. (2015). Pragmatism, moral responsibility or policy change: The Syrian refugee crisis and selective humanitarianism in the Turkish refugee regime. *Comparative Migration Studies*, 4(2), pp. 2–20.

Marfleet, P. (2007). Refugees and history: Why we must address the past. *Refugee Survey Quarterly*, 26(3), pp. 136–148.

Marfleet, P. (2013). Explorations in a Foreign land: States, refugees, and the problem of history. *Refugee Survey Quarterly*, 32(2), pp. 14–34.

Özcan, A. (1997). *Pan-Islamism: Indian Muslims, the Ottomans and Britain* (1877–1924). Leiden: Brill.

Pérouse, J-F. (2013). La Turquie face aux soubresauts migratoires dans un contexte de crise. *Confluences Méditerrannée*, 4(87), pp. 85–93.

Sabah (2018a). Cumhurbaşkanı Erdoğan: Abdülhamid Han Dünyanın Son Evrensel İmparatorudur. 10 Feb.

Sabah (2018b). Cumhurbaşkanı Erdoğan: Mehmet Akif'i Geleceğe Taşımalıyız. 26 Feb.

Togral-Koca, B. (2016). Syrian refugees in Turkey: From 'guests' to 'enemies'? *New Perspectives on Turkey*, 54, pp. 55–75.

Umar, Ö.O. (2003). *Türkiye-Suriye İlişkileri (1918–1940)*. Elazig: Firat University Middle Eastern Research Center.

Walters, W. (2006). Security, territory, metagovernance: Critical notes on antiillegal immigration programmes in the European union. Conference Paper presented at Istanbul Bilgi University, 7 Dec.

Werz, M., Hoffman, M. and Bhaskar, M. (2015). Previewing Turkey's general election, 2 June. Available at: https://www.americanprogress.org/issues/security/report/2015/06/02/114206/previewing-turkeys-general-election/ [Accessed 5 Oct. 2016].

Wodak, R. (2010). *The discourses of politics in action: Politics as usual*. London: Palgrave.

Wodak, R. and van Dijk, T.A., eds. (2000). *Racism at the top: Parliamentary discourses on ethnic issues in six European states*. Austria, Klagenfurt/Celovec: Drava.

Part III

Everyday spaces of tolerance

8 Paradoxical visibility

Purpose-built Mosques in Copenhagen

Kirsten Simonsen, Maja de Neergaard and Lasse Koefoed

> *Last Friday Muslims all over the world prayed for Hvidovre's mosque. Guests from eleven European countries headed by the leader of Islam opened the first mosque in Scandinavia and with that made Hvidovre the Mecca of the North.*
>
> (Hvidovre Avis, July 28, 1967)

The epigraph is the headline in the local newspaper on the occasion of the inauguration of the *Nusrat Djahan Mosque*, a small mosque purpose-built in a suburb of Copenhagen in 1967. A speech from the mayor followed: "We in the municipality can only express delight at the mosque being built here. Religious divisions can sometimes bring emotions to a boil, but in this country we are fortunately very tolerant. Therefore we welcome Islam and the new mosque in Hvidovre!". At the national level, the opening of the mosque did not draw much attention, and the one it received was characterised by openness and curiosity.

A lot of water has flowed under the bridge since then, and we can, with the words of Göle (2011), speak of a 'loss of innocence'. Mosques are no longer simply sites confined to the needs of the pious and of local inhabitants; they draw both political attention and public debate. This exposure is, as has been highlighted in the introduction to this volume, part of a wider visibilisation granted to any and all issues concerning Islam across Europe. Islam is, indeed, no longer an external reference but also an internal one, part of wider debates surrounding European self-awareness, identity and politics. In this process, conflicts over the building of mosques can be seen as negotiations over a politics of identity, fought over architectural form and location. Purpose-built mosques with domes and minarets have become, by now, seen as representative of the presence of Islamic cultural heritage in Europe. The dome and the minaret have become a 'structural metonym' of Muslim identity, and it is in particular these visible elements that have generated conflicts many places (see e.g. *Journal of Ethnic and Migration Studies*, 2006). These religious buildings have indeed take on an important affective power: capable of inspiring feelings of emotional security and sanctuary among their congregation but also feelings of insecurity, anxiety, and hostility among opponents.

When it comes to the building of purpose-built mosques, apart from the small one introduced earlier, Denmark is a late-comer in Europe. In the region of

Figure 8.1 Nusrat Djahan Mosque, Copenhagen

Copenhagen (as the only place in Denmark) we now have three purpose-built mosques. The two other ones, considerably bigger than the first one, are built in the central part of Copenhagen, not so far from each other, in one of the city's most multicultural districts. First, the *Khayr El-Bareya Mosque* with a connected community centre was inaugurated in 2014, and it was followed by the *Imam Ali Mosque* in 2015.

This chapter will focus on the reception of these two edifices, putting the main emphasis on the *Khayr El-Bareya Mosque* as it was the first one to be built, and because we have carried out the most intensive fieldwork around its opening. The chapter will be structured in four parts: first, we will look at the Danish context, that is, the discursive climate of the public debate around Islam as it has developed from the positive curiosity illustrated earlier, towards an imagination of 'the Muslim' as a threatening figure; second, we will present our take on the issue of in/visibility and the paradoxes involved in the concept; third, we will analyse the dominant reception of the mosque in politics and in the media; and finally, under the heading of spaces of hope, we will present and discuss where we perceive cracks and paradoxes in the dominant discourse.

The Danish context

Even though neither immigration nor negative attitudes toward immigrants are new phenomena in Denmark (see e.g. Simonsen, 2015), the common imagination is that the country was culturally homogeneous until the 'new' immigration

Figure 8.2 Khayr El-Bareya Mosque, Copenhagen

Figure 8.3 Imam Ali Mosque, Copenhagen

began in the late 1960s. This immigration took two forms. First, groups of 'guest workers' arrived, mainly from Turkey, Morocco, Yugoslavia, and Pakistan, later followed by their spouses and children who came by way of the right to family reunion. Second, from the middle of the 1980s, immigration took mostly the form of political asylum given to refugees from countries such as Iran, Iraq, Palestine,

Bosnia, Somalia, and, as the latest, Syria. However, it was not until the late 1980s that the debate on Islam and cultural difference appeared in earnest. Yilmaz for instance writes, "I went to Denmark in 1979 as a young, left-wing activist who had no idea what the term "identity" meant. Within two decades, I became a "Muslim"" (2016, p. 3), marking a personal experience of this discursive development. Becoming a 'Muslim' in this way is connected to a growing imagination of Islam as a threat.

As the Western world was shaken by a series of terror attacks, stretching from the 9/11 attacks in the United States, the London Underground bombings, and, amongst others, the attacks on *Charlie Hebdo* magazine and Jewish supermarkets in Paris, to those at the public event 'Art, Blasphemy and Freedom of Expression' and at the Synagogue in Copenhagen. These attacks have given occasion for the construction of Muslims and their actions in the name of religion as a fundamental threat not only to citizens *in* the West but also to the basic freedoms and values *of* the 'West'.

The moral panic in relation to immigration has however, not been limited to the fear of terror. In the Danish public debate we can identify three other forms of threat among which the earliest one is a *threat to the welfare state*. This framing represents the Danes as a tolerant and naïve people trying to do good for everybody but merely opening up themselves to exploitation. An illustrative example is a campaign run by a Danish tabloid paper in 1995. It was labelled 'The Price of Goodness' and aimed to illustrate through examples how immigrants received disproportional help from public funds compared to 'Danes' in social need, and how this was what attracted immigrants to Denmark (see also Koefoed and Simonsen, 2007).

The second one is the *threat from demography*. On several occasions, Danish newspapers have connected the 'immigrant problem' to statistical extrapolations. One example is a series from the right-wing paper *Jyllandsposten* called 'The new Denmark' based on a public population forecast. It claims that immigrants from non-Western countries during the next 100 years will increase by 350%, summing up their conclusions by writing that "mousy hair and blue eyes will no longer signal that you are looking at an average Dane; it might as well be black hair and brown eyes" (*Jyllandsposten*, July 15, 2001, here cited from Hervik, 2002, p. 164). The article expressed anxiety about the risk that the Danes in a few decades will no longer be a homogeneous population with roots in the same religion, culture, traditions and norms. Such a discourse appeals to (cultural) racism naturalising and essentialising cultural difference. The premise is that immigrants (and in particular Muslims) are bound to cultures that are alien to the Danish one and resist integration into the Danish society. As such, it constructs a rigid dichotomy between 'us' and 'them' where the differences are seen as insurmountable and inalterable. Danish society and Danish culture are thus considered threatened by immigrants 'polluting' them with their strange norms and values.

These values are described as "the forces of darkness", "medieval religious values", or "obscurantism and black Middle Ages" in this way designating a *threat from the Middle Ages* (Hervik, 2002). The contradiction between 'us' and 'them' is thus also formulated in terms of modernity versus tradition by way of relating to issues such as human rights, enlightenment, religion, and not least gender relations. Gender equality is represented as something *we* have and *they* do not, 'they' being (Muslim) immigrants deeply embedded in their 'medieval' culture. The figure of the Muslim stands for female suppression, fundamentalism, violence, terrorism, and cultural dissension. *He* is a traditionalist and difficult, if not impossible, to 'modernise'. The image of a threat from the Middle Ages thus involves a rhetoric of *time-space separation*, a positioning of 'the other' in a distant place and a distant time. At the same time it involves an internal *territorial containment*, signifying the so-called ghettoes in the big cities as spaces of failed 'integration' and need of restrictions.

The indisputably most conspicuous media event in Denmark in connection with these debates was the publication in 2005 of the Mohamed drawings in *Morgenavisen Jyllandsposten*. It represents a definitive retreat from any semblance of multicultural politesse. This turned out to be 'a long and messy event' (Hervik, Eide and Kunelius, 2008) primarily unfolding in 2005/06, but with aftershocks and subplots continuing at least into 2010.[1] The original publication was narrated as an editorial decision to draw attention to assumed self-censorship reported by a Danish author complaining about difficulties in finding an illustrator for his children's book on Islam. As a response to this story, the editors asked a number of satirical cartoonists to draw caricatures of the Prophet Mohammed 'as they saw him'. The 12 drawings were published on September 30 of that year, accompanied by a text saying that Muslims, when living in a modern, secular society exercising democracy and freedoms of speech, cannot expect consideration of their religious feelings. They "must be ready to put up with insults, mockery and ridicule". Initially the publication provoked little debate; probably because *Jyllandsposten* (as illustrated earlier) just followed a previous pattern with provocations and Islamophobic articles (see e.g. Hussain, 2000; Hervik, 2002; Andreassen, 2007). Only with growing reactions from Muslim groups and others in early October did the media interest grow. When the Danish Prime Minister refused to meet 11 ambassadors from Muslim-majority nations who asked for a meeting to discuss the situation of Muslims in Denmark, the basis was laid for a globalisation of the controversy. The result was the well-known story of demonstrations, economic boycott activities, and violent attacks on Danish missions abroad.

Hervik (2008) has analysed the framing of the event, and he identified three broad frames employed: (1) "Free speech is the issue, and it is a Danish issue", (2) "freedom of speech as an universal human right, with limitations", and (3) "demonization of Muslims is the issue, not free speech". The first frame became the dominant one in the domestic news coverage, not least promoted by spin from the Danish government and neoconservative connections across the Atlantic (Hervik, 2008). The dominant view was one of a 'clash of civilisations' between Islam and 'the Western world', and of free speech as a Danish and Western 'property' under threat.

In the public debate in Denmark there has also been an obsession with Muslim clothing. Questions surrounding the 'hijab', 'niqab', and 'burka' have been some of the most controversial issues in relation to religious pluralism. Maybe one of the most absurd controversies was a heated public debate on burkas running in Denmark during 2007. It initiated a publicly commissioned investigation that, at the end, showed a near-total absence of burka-wearing in the country. In these controversies over girls' and women's dress a range of liberal principles were regularly activated: state neutrality, gender equality, religious freedom, multicultural accommodation.

The most common framing of these debates has been the *victimisation* frame (Andreassen and Lettinga, 2012). Muslim women are represented as oppressed by their community and in need of liberation. The arguments are that headscarves and veils conflict with the 'Danish value' of gender equality, that they are gender-specific types of clothing which mark a hierarchy between men and women, and that they are a structural problem embedded in Islam. When Muslim women do not themselves consider the veil oppressive, it is viewed as false consciousness due to their internalisation of gender discriminatory norms. Inspired by the Norwegian dramatist Henrik Ibsen, Lentin and Titley (2011) talk about how the European veil-debates have 'doll-housed' Muslim women, failing to note how fixation on veils positions these pieces of clothing as a racialised signifier of civilisational hierarchy.

Some populist fractions of feminism have gone beyond the victimisation frame. For example Thomsen (2000), in a newspaper feature "Whore and Madonna again", blamed veil-wearing women for promoting sexual assaults on 'Danish' women. She connects such assaults to the gendered practising of Islam: Muslim men become the violating ones and Muslim women, because of their religious markers as 'pure women', are claimed to legitimise men's violations. Wearing a scarf, she argues, desexualises the wearer while at the same time sexualising non-wearers and rendering them open for assaults. This angle of seeing the veil as *offensive* to the 'Danish' majority becomes even more obvious in the media storm that has met young independent women with headscarves taking public positions, or running for parliament. They have been opposed with statements such as "Women for Freedom takes distance from the hostile use of the religious headscarf" or protests against their candidacy such as the one from a MP for the Danish People's Party comparing Islam with Nazism and the headscarf with the Nazi Swastika.

Taking into account the at times infamous debate in Denmark and the strength of a populist right-wing party, it seems paradoxical that the country has not, like some other European countries, imposed restrictive measures against veiling. The Danish People's Party in 2004 proposed a ban on "culturally determined head coverings" of public employees. The proposal was, however, rejected by all other political parties, and the Minister of Integration (coming from a liberal party) declared that he was not Minister "of headscarves and liver pâté". Hadj-Abdou et al. (2012) explain that result with a strong notion of individual rights (including religious freedom), which is deeply ingrained in Danish political culture, combined with an understanding of religion as a private matter even when expressed in public space. In this way the 'defence of liberal values' can take a double role both as a strategy against Islam and a limit to populism. However, since this analysis was conducted, we have experienced a further radicalisation of Danish politics on 'strangers', and on August 1, 2018 a burka ban was introduced under the pretext of a masking ban, since religious discrimination is against the Danish constitution.

Many of the issues touched on in this section can be summarised in what Tebble (2006) calls *identity liberalism*. Here, the idea of liberalism changes from describing a *modus vivendi* between many different ways of life, to be seen as a way of life in itself – which easily turns into a mechanism of exclusion (as Brown, 2006 has argued). The proponents advocate a national identity of shared liberal values and the duty of the state to protect the 'open society' from the illiberal forms of life of 'others' – often by way of illiberal precautions. It is in this framework that the opening of the big mosques in the central part of Copenhagen shall be understood.

In/visibility

The framing described in this chapter draws the attention to the question of *visibility* of Islam in public space – and so does the issue of purpose-built mosques. In the present analysis, the notion of visibility is embedded in an overall understanding of cross-cultural embodied encounters. Briefly outlined, this understanding is theoretically based on a combination of third-generation phenomenology, practice theory, and postcolonialism that takes inspiration from Merleau-Ponty's phenomenology of the lived body and emphasises the interrelational structure of our embodied existence (see e.g. Simonsen, 2007; Koefoed and Simonsen, 2010). In this understanding, the embodied encounters set an ontological interdependence between seeing and being seen – it is our own visibility, our embodiment in the world, that grounds the possibility of our seeing. In the present connection, however, the seeing/being-seen and the cross-cultural encounters take a specific form as they are mediated through the materiality, the architecture, and the symbolic signs of the purpose-built mosques. Such encounters, following Merleau-Ponty, involve social bodies sensing and intertwining with the world (and its materialities) in a way that belongs neither to subject nor object but produces attunements and emotions in their interspace. As material and visible signs, the mosques become interfaces among citizens of different confessions or secularities.

The notion of visibility has been employed in a variety of ways, not least in studies of migrants and ethnic minorities (for a recent discussion see the special issue of the *Nordic Journal of Migration Research* 2014). Brighenti (2007) by definition nominates issues of recognition and control as being the two most important social outcomes of visibility. At the same time, she underlines their oppositional character, referring to the ambivalences of visibility and their effects. In this instance, we will emphasise the aspect of recognition, basically following Honneth (1995) in his suggestion that intersubjective recognition stands as a necessary condition for intact identity-formation.

For a useful take on the relationship between visibility and *identity* one might look to Linda Alcoff (2006) who explores race and gender as visible identities and describes some of the mechanisms by which they are identified, enacted, and reproduced. Her approach is mainly phenomenological and emphasises how social identities are transcultural, historically fluid, and grounded in social contexts. These social identities, she argues, imply a recognition of bodily difference. We might imagine subjectivity as mind and imagination, but the social identities of race and gender operate ineluctably through their bodily markers: "they do not transcend their physical manifestation because they *are* their physical manifestation, despite the fact that the same features can support variable identities depending of how the system of marking works in a given culture" (2006, p. 102). Hence, to a large degree, the constitution of social identities occurs through negotiations in public space, over mobility and home, (imagined) communities, boundaries and bridges, rights to urban space, etc. Part of that process is the negotiation of difference. 'Like bodies' and 'unlike bodies' do not precede encounters of inclusion or expulsion. Rather, likeness and unlikeness as 'characteristics' of bodies are produced through these encounters – a process where oppressive visions are marking the 'unlike' bodies (cf. Ahmed, 2000). Fanon (1967) describes the phenomenology of incorporating Otherness and the development of a 'double consciousness' due to the enculturation of the body. Men and women of colour, he says, develop a third-person consciousness trying to reconcile their own experiences with the operation of a 'historical-racial schema' within which their corporeal schema is supposed to fit. In this way, he is arguing that for all coloured people in the 'colonial' world, it is the self-consciousness about being a body-for-others that dominates their consciousness in public settings.

The relationship between the visibility represented by the mosques and the formation of identities emerges clearly from interviews made with different participants in the opening of the *Khayr El-Bareya Mosque*. In the joy expressed by the participants, the interviews included remarks such as "this is a historical and a very special experience for me", "it is really great to come to a place where you feel at home", and "now, I can better feel at home and call myself a Danish Muslim". It is obvious from these remarks that this is not only about identity and subjectivity but also about politics, in this way making topical an Arendtian politics of in/visibility (Arendt, 1958; Borren, 2008). Arendt defines the political world as the *space of appearances*, "the space where I appear to others as others appear to me" (1958, pp. 198–199) in this way closely relating it to visibility. From her reflections on the condition of the stateless, she distinguishes between the two conceptual pairs: the first pair is 'public visibility' and 'natural invisibility' that she considers capable of constituting the criteria of sound political action and citizenship; the other one is 'public invisibility' and 'natural visibility' which is politically harmful and disabling since the individuals here are reduced to pure organic life, to the naked naturalness of being-nothing-but-human. The natural visibility then comes close to Fanon's thinking on visual exposure and the 'historical-racial schema'.

Now, racial identity as well as race and racism are contested terms in the European/Danish contexts (e.g. Goldberg, 2009; Bech and Necef, 2012; Rabo and Andreassen, 2014; Leinonen and Toivanen, 2014). Goldberg (2009) has convincingly argued how the history of Holocaust has rendered the notion of race unspeakable and invisible, but also how the notion is 'buried alive' (2009, p. 157).

The denial of (biological) race does not erase racism. Racism is not simply a matter of biology, it has always also been cultural. Race has always had to do with sets of views and predispositions concerning culture, or culture tied to colour. The most conspicuous example is here the current imagination and role of 'the figure of the Muslim' in many European countries. In the dominant European imagination, Islam seems to represent a collection of 'lacks': of freedom, of civility and manners, of love of life, and of equal respect for women and gay people. 'The Muslim', as also seen earlier, is framed as a threat that ferments violence, militancy, terrorism, and cultural dissent. One element of this imagination has been a racialisation of all visible signs of Muslim religious affiliation such as long beards, Muslim clothing, female headscarves, and buildings with Islamic characteristics.

Such processes have fruitfully been explored in recent ideas of *whiteness* (e.g. Dyer, 1997; Ahmed, 2004, 2006; Shaw, 2007, and in the Nordic context, Loftsdottir and Jensen, 2012) that emphasises how the notion is not just a characteristic of bodies; it is an ordering device getting its power by becoming habitual. Whiteness is invisible and unmarked, as an absent centre against which other groups appear only as deviants. We do not see white bodies as *white* bodies, just as bodies. This normalisation renders possible to talk about 'white spaces' in the sense that spaces are orientated around whiteness through the repetition of acts, allowing passing of some bodies and not others (Ahmed, 2006). And yet, non-white bodies do inhabit white spaces. However, the theory says, such bodies often are *made invisible* when spaces are seen as white. At the same time non-white people can become *hypervisible* when they are not capable of passing, they 'stand out' and are seen as 'out-of-place'. Similar conceptualisations could be materialised and applied to the mosque-case when we consider the transformation from indistinguishable and invisible prayer rooms to purpose-built mosques including 'hypervisible' Islamic signs such as domes and minarets. The concept of whiteness gives an important contribution, also to the understanding of Danish society more broadly (Simonsen, 2015). In practise, however, social relations of in/visibility are more complex (Leinonen and Toivanen, 2014). Brighenti (2007) characterises visibility as a double-edged sword – it can be empowering as well as disempowering. In this sense, there is a built-in ambivalence in the notion of in/visibility, in particular when they refer to issues of identity and recognition. The term 'hypervisibility', for example, interprets visibility as an undesirable condition imposing cultural and racial stereotypes on minority groups. But it is also part of 'politics of identity' claiming recognition in public space through visible manifestations of difference. Similarly, invisibility can relate to the powerlessness of the overlooked and unnoticed, as well as the possibility of passing or even strategic performance of successful integration (see e.g. Juul, 2014).

This in/visibility, with its load of identity and ambivalence, is also significant when it comes to material form, architecture, and aesthetics. Göle (2011) writes:

The 'visibility' of Islam in public is [. . .] a form of agency, a manifestation of religious difference that cannot be thought independent of the materiality of culture, namely aesthetic forms, dress codes, or architectural genres.

(2011, p. 383)

And this agency takes the form of negotiations and politics of identity fought over design, architectural form, and location. That is what we saw when the man cited earlier felt that the material manifestation of Islam in the new mosque at long last made it possible for him to call himself a 'Danish Muslim'. As Göle also has noted: "one becomes a citizen as one makes oneself visible to others" (2011, 390).

Such material transformations have happened across Europe by now, though much later in Denmark than in many other European countries. Until the recent opening events, mosques have mostly been discrete and undistinguishable. In this sense, strategies of invisibility, as much as visibility, have been working among Muslims, and the recent openings have represented a new strategy of visibility – even though they are embedded in a series of former attempts and failures. This visibility also becomes paradoxical as it potentially facilitates both community and distance. Depending on the specific contexts, the encounters circumscribing the mosques simultaneously contain agreement and conflict. It is this paradox that we wish to emphasise by the term 'hypervisibility'. In the present-day Danish context, the visibility of purpose-built mosques has proved to be significant (i.e. 'hyper') since their construction and opening has evoked compounded and contested reactions.

Architectural form is an important element here. Everywhere purpose-built mosques with domes and minarets represent the imprint of Islamic cultural heritage. As we noted previously, the dome and the minaret have become a 'structural metonym' of Muslim identity, also generating conflicts, in particular when an audible call to prayer from the minarets is added (Gale, 2005; Göle, 2011). In the Danish case, the three mosques in very different ways express adaptations of this 'other' type of building to the Nordic context, and in all cases this involved a specific attention to the minarets. In the early, suburban mosque, the architect and the congregation by themselves decided to refrain from building a minaret since they judged it out of proportion with the surrounding buildings. Both of the two new big central mosques have minarets, but in the negotiations with the municipality (and due to extensive objections) the result was a compromise securing that the minarets remain silent.

To recapitulate, the question of visibility of Islam in public space is constituted by a plurality of perspectives. It is full of paradoxes and it gives occasion for action, conflict, and confrontation. From the previous, we want to stress first the importance of material signs and symbols; second, the performative practices by which groups and individuals make their differences and appearances explicit to others; and third, the cultural struggles over memory and visibility in the public sphere. In the following section, we examine by way of illustration the public debate around the opening of the *Khayr El-Bareya Mosque* in Copenhagen.

Political and media reactions

The opening of the first purpose-built Mosque *Khayr El-Bareya Mosque* in June 2014 was followed closely by leading Danish media and created heated debate on this new visibility of Islam in the centre of Copenhagen.

The main circulating headline in the electronic media the day before the opening event was '*The official Denmark opts out*'. It was about the Royal Family, the Government, and party leaders who refused to participate in the opening event. The most dominant voices of opposition came in reaction to a big sponsorship deal with the mosque (from Qatari Emirates) and to a newspaper interview where a mosque spokesperson described homosexuality as *haram* (forbidden) and an illness. The Minister of Integration "could not find room in his diary", and a left-wing party leader was "not religious" and would therefore not attend the opening of a mosque. When it comes to the municipality of Copenhagen, The Chief Burgomeister also used the excuse of a busy diary, but he compromised by sending the Mayor for Social Affairs who is also a Social Democrat. A few politicians – interestingly from parties of different political orientation – argued differently, expressing that if you want dialogue with Danish Muslims, you have to turn up and engage. In the end, two politicians participated, one from the Christian Democrats and the other from the Socialist People's Party. The latter balanced her appearance (and made her opinions clear) by wearing a LGBT rainbow badge during the event. From the municipality only three out of eight mayors turned up. So, the political elite were sparsely represented, as were the leaders of trendsetting media. In this sense, the dominant secular power regime in Danish society managed to set its mark on the event from the outset.

Tolerance and stigmatisation

There are good reasons for congratulating the Danish Muslims with the new mosque with both minaret and dome. It is not only good news for Danish Muslims but also for the Danish broadmindedness, freedom of religion and pluralism.

(Stormoské June 20, 2014)

Everyone has the right to practice their religion in Denmark. The problem arises if there are hidden ideologies behind the neat exterior.

(Kadher June 21, 2014)

The new mosque is a symbol of freedom of religion.

(Cekic June 20, 2014)

One framing of the public debate has a clear reference to the *symbolic* discussion of what can be accepted and tolerated in public space. The mosque here serves as a religious symbol for something else; it is not about the mosque in itself but about the significance and interpretation of it. In this case the appearance of the mosque is interpreted as a symbol of 'our' values: freedom of religion, democracy and openness. The logic is that a liberal secular state such as Denmark should accept the visible presence of Islamic religious symbols in public space (Lægaard, 2010 and his chapter in this volume). But as we will argue, it works as a special kind of acceptance, not respect or recognition but rather *tolerance* with the negative connotations this word has. The term is signalling a will to accept the presence of something for what you in reality have aversive feelings towards.

More generally, we, together with Brown (2006) can understand tolerance as a political discourse, regime or order of politics that involves a kind of depolitisation or culturalisation. She argues:

> Tolerance as such is not the problem. Rather, the call for tolerance, the invocation of tolerance, and the attempt to instantiate tolerance are all signs of identity production and identity management in the contexts of orders of stratification or marginalization in which the production, the management, and the context themselves are disavowed. In short, they are signs of a buried order of politics.
>
> (2006, p. 14)

In this case tolerance is not about a personal ethics but rather about the way tolerance is produced with social and political effects. One of the important framings of the public debate is the promotion of tolerance where it is presumed that 'we' are always already tolerant being a democratic, civilised, and secular society proven by the very fact that we accept and tolerate the new materialisation of Islam in the form of a Mosque in the middle of Copenhagen. But the debate also becomes characterised by reservations, anxiety, and uncertainty and dominated by statements like: "Is there something hidden inside the Mosque?". For example, an editorial from the newspaper *Politiken* that is considered culturally progressive, under the headline "the Mosque is welcome without Qatar", represents the donation from Qatar as something that 'we' should fear because it can potentially affect the orientation of the Mosque. And other voices proceed along that line:

> But the joy is tempered by disappointment that no private donations have been granted in Denmark. Instead it is financed from Qatar – a country that is very far from open and tolerant and is known for its support to Islamists in Tunisia, Libya, Syria and Egypt. Many fear that the orthodox and conservative interpretation of Islam in Qatar will cast shadows all the way to Rovsinggade.
>
> (Stormoské June 20, 2014)

> Mosques are unavoidable in a democratic society but dangerous. It has to be closely monitored 'that nothing is taught or done that is against good morals or public order'. [. . .] Because Danish culture is incompatible with Islam and Islam is incompatible with Danish culture.
>
> (Hvorslev, June 11, 2014)

Now, the mosque symbolises 'them' – in sharp contrast to 'us' – lacking 'our' values such as freedom, tolerance, and openness. Thereby tolerance is increasingly sliding into stereotypical us-them binaries that essentialise the difference between 'Danishness' and Islam by culturalising and naturalising the difference.

Securitisation and the enemy of tolerance

On the day of the opening, the Mosque was welcomed by a satirical drawing by *Berlingske Internet Media*, leading the thoughts back to Denmark's earlier media cartoons (Figure 8.4).

This cartoon draws the attention to another dominant issue of the debate, that is, the *security* issue which is mainly concerned with the mosque potentially becoming a place for radicalisation and terror plots. Here the Mosque becomes the *enemy of tolerance*. As Brown argues, "in the aftermath of September 11, political rhetorics of Islam, nationalism, fundamentalism, culture and civilization have re-framed even domestic discourses of tolerance" (2006, p. 6). The threat and security issue was very dominant in the public debate, constructing the mosque as a dangerous place, one outside the boundaries of civilisation and thus a threat to Danish society. The securitisation of the mosque was expressed in an editorial in

Figure 8.4 Cartoon from Berlingske Internet Media

the conservative newspaper *Jyllandsposten* linking the mosque to the war in Syria and Islamic fundamentalism:

> Muslim worship places have been and are reasonably believed to be recruiting centres for Islamist radicalism, which among other things has led young people to take part in the so-called holy war in Syria. It also moderates the enthusiasm that the Danish Islamic Council is believed to have close links to the strongly conservative Muslim Brotherhood. So altogether we must supplement our welcome with the wish that the Security Intelligence Service will follow closely the life and deeds of the mosque.
>
> (Dæmpet velkomst June 19, 2014)

The construction of the mosque as a site of danger, terror, and war followed a predictable script. It immediately puts *imaginative geography* to work, constructing the mosque as something invading Danish territory and relating to a geopolitical logic of *everywhere war* and the *colonial present* (Gregory, 2011). The mosque not only brings Islam into the cityscape but also the war on terror. In this way it is an imagined threat to society, linked to recruitment for Islamic radicalism, terror, and violence. The mosque was not only linked geographically to Qatar but also to Palestine and Syria, inhabited by mythical and dangerous figures like Hamas and ISIS. Similar negative comments came, for example, from a central MP from a (neo)liberal party: "I hope that the Security Intelligence Service will keep an eye on them" she said to *Jyllandsposten* (Johansen, June 15, 2014), thereby linking the Mosque to the new order of military urbanism (Graham, 2011). The Mosque becomes synonymous with the intolerable and aggressive subject, that threatens our tolerance and civilisation. This is an example of how the Mosque becomes securitised. The Mosque is welcomed by installing a point of no return that includes the legimation of a new form of state action. In this way, the local discussion was put into an civilisational discourse that construes an opposition between cosmopolitan West and its putatively fundamentalist 'Other'.

Another part of the debate represented in the cartoon in Figure 8.4 with the little dustbin with the inscription "for gay and lesbianin infidels" was the issue of *sexuality and gender*. The gender issue was mainly taken up in relation to the physical separation of men and women in the mosque. The mosque and its material and architectural design were taken to stand for female suppression and traditionalist views on gender that are impossible to 'modernise'. In this way, gender equality was represented as something *we* have and *they* do not, 'they' being Muslim believers deeply embedded in their backward culture. It followed what we earlier have described as the 'victimisation frame' representing Muslim women as oppressed by their community and in need of liberation. Sexuality became actualised in the debate due to the earlier mentioned newspaper interview where the former spokesman for the mosque described homosexuality as *haram*. The interview drew strong reactions in the media and came to add to the representation of Islam as homophobic and an enemy of tolerance.

The drawing thus captured the many different dimensions of the debate: the dominant imagination of the other as representing a collection of lacks: of freedom, civility, and equal respect for women and gay people, here, visualised and racialised in a satirical way, connecting it to militancy and terrorism. The cartoon also immediately created historical links to the former cartoon crisis in Denmark in 2005/6. A media event and crisis that turned out to be a long and 'messy' event where the dominant view was, as we previously argued, a clash of civilisations and a situation where Danish free speech as a specific western property was under threat.

Spaces of hope

We began this chapter by highlighting the openness and positive curiosity with which the *Nusrat Djahan Mosque* was met on its opening in 1967. In that moment, the opening of a mosque was received as an *exotic* element amidst the 'normality' of the Copenhagen suburb. During the 50 years that have passed since, the mosque has managed to function in relatively unquestioned and undisturbed fashion. It has, you could say, developed from being an exotic stranger to becoming a familiar stranger. In contrast, the two bigger mosques in the central part of the city opened in the present-day context which, as we have seen, has been heavily marked by Islamophobia and fear of terror. In the current political climate, we have experienced a profound *politicisation* of issues around 'other' sites of worship, in particular when it comes to Muslim sites. Today, it is all about the public visibility of Islam. Visibility manifests Muslim politics of identity, taking the form of religious practices that are embodied and spatial and performed in specific material and aesthetic forms. The concept of visibility also encompasses personal identification as well as public manifestations of Islamic difference. An illustrative example is the man for whom the possibility of worship in a 'real' mosque meant that he could now, with confidence, call himself a 'Danish Muslim'. All of this is condensed in the movement of the mosques from invisibility to visibility and the public reception of these moves. While opposition in public space and the media was manifest, paradoxes did, nevertheless, arise around the opening events revealing cracks in the dominant reception and opening spaces of hope.

One obvious paradox arises from the difference in attitude to participation in the opening ceremony between the political and the clerical establishment. While the majority of the politicians, both left- and right-wing, excused themselves on more or less reasonable grounds, the different religious communities were present to welcome the 'newcomers'. At the opening of the *Khayr El-Bareya Mosque* a range of religious representatives participated, leaving the impression that religious communities recognise each other's existence. The Bishop of Copenhagen was represented, and the vicars from the neighbouring churches were there to congratulate their Muslim brethren. The leader of the Catholic Church was there, and so were leaders of the Shia Muslim community. In this sense the ambition of the mosque to open a dialogue between different religious communities was successful. A representative of the Bishop of Copenhagen gave a speech at the inauguration. He congratulated the new mosque and said: "We are delighted for you that you have got this new and beautiful place. It is obviously inspired by Nordic architecture and gives it a new beautiful expression. It unites the Nordic with the traditional Islamic" [our translation]. In this way, he touched on the visibility and recognised the efforts of the community to build a mosque expressing a 'Nordic' architectural form.

The invited guests at the opening of *Imam Ali Mosque* were primarily Middle Eastern imams, representatives from local religious communities of all kinds, and local NGOs. The inauguration ceremony culminated with the symbolic act of cutting a ribbon to the central part of the building. This act was performed by a local vicar from the Danish state church, allowing the organisers to underline, in this fashion, the local orientation of the mosque and the will to religious dialogue. Another part of the opening took the form of a mini-conference with talks from a number of representatives and invited speakers. The programme was international in its outreach, but at the same time it was clear that the imam and the people around the mosque were highly integrated with the local community. Many participants in the ceremony came from local religious organisations. The mosque has existed in that location for more than a decade, and the establishment and opening of the purpose-built mosque was the culmination of a long-hoped dream and a lot of hard work.

We experienced a similar mutual will to religious dialogue in another case study on cross-cultural encounters, focussing on multicultural festivals (with also many of the same individuals involved). At a festival called *Smag verden* ('Taste the world') many of the activities were organised dialogically in the way the different performances were actively involving the participants. Amongst them was a '*tent of faiths*' where four religions were presented, Christianity, Judaism, Islam, and Buddhism respectively. The atmosphere in the tent was bustling with activities. The four representatives were standing each by a high café table inviting people to ask questions on religious issues, in this way performing a direct religious dialogue with visitors. People were invited to participate in a religious quiz. By being present together in the tent they created a feeling of common interreligious engagement. This is an example of religious encounter that works with the elimination of barriers among people created by different forms of stereotypes and hierarchies, replacing them with more equal and respectful relations. The atmosphere in the tent was related to the attempt to define meaningful concepts of religious identity and belonging, rather than violent confrontation and condemnation.

Another paradox appears when we relate the public media debate to the reactions in the immediate surroundings of the mosques. Here we saw everyday spaces of tolerance in the form of a positive atmosphere and interested curiosity circumscribing the openings of the two recent mosques, in particular by visitors from the local neighbourhoods. The positive curiosity is visible in the on-the-spot interviews that formed part of our fieldwork on the *Khayr El-Bareya Mosque* (the fieldwork on the *Imam Ali Mosque* was designed differently). Certainly, the interviews demonstrate variety as some responses were pessimistic or ambivalent, but the majority of the reactions were optimistic and inviting. Among this latter, dominant, group some responses expressed genuine joy – either on their own behalf or on behalf of 'their Muslim neighbours'. Others expressed how the opening of the *Khayr El-Bareya Mosque* is significant because it "finally provides a proper place for worship" to the Muslims and that this is important for Danish society in general. On the question of how they envisage the mosque will influence the local community, it is possible to distinguish the positive responses between those visitors who came from outside the neighbourhood (i.e. from other parts of the city and from outside the city) and those who are locals and live nearby the mosque. Among the visitors who live outside the neighbourhood and came from other parts of the city/region their responses emphasised how it is "good for the Muslims to have a common ground" and that the *Khayr El-Bareya Mosque* will enable the Muslim community to signal greater 'openness'. The responses from the locals showed more detailed notions of how they expect the mosque will be able to play a role in the local community. Some expected the mosque to make the neighbourhood 'more lively' and to be able to involve some of the local youth in different activities and 'help keep them off the streets'. To the dominant positive group, the general motive for participating in the event was to show respect and tolerance: as one man stated: "I am mainly here to express sympathy and inter-est with this newly started project. [. . .] They start out in a positive manner with great openness. If only the Danes would realise that, I believe this openness will matter." (Male, age 45). Participating in the opening event can also be the result of a continuous reflexivity, as a woman states: "I have a number of religious girl-friends so I want to see what it is that means so much to them – that they now have a proper mosque". When asked about her perception of the public debate she said:

> Well, it means that you need to argue why you choose to come. So I have to reflect on why I'm doing it. And I think that no matter what, and no matter all sorts of different attitudes, I believe that the most important thing is that people living in our society have a sense of having a place where they feel accepted and able to practice their belief.
>
> (Woman, age 25)

In sum, while newspapers and public debates emphasised resistance and con-troversy towards the two recent, purpose-built mosques, our research has high-lighted that the general responses on-the-ground during the opening events were manifest with tolerance and spaces of hope. In this sense, the direct and mediated encounters performed in relation to the new visibility of Islam in the public space of Copenhagen underline the paradoxical character of in/visibility, with its load of identity and ambivalence, and hospitality as well as hostility.

Note

1 In January 2010, an attempt was made on the life of Kurt Westergaard, the designer of the most aggressive of the Mohammed cartoons, depicting the prophet with a bomb as a turban.

References

Ahmed, S. (2000). *Strange encounters: Embodied others in post-coloniality.* London and New York: Routledge.

Ahmed, S. (2004). Declarations of whiteness: The non-performativity of anti-racism. *Borderlands E-journal*, 3, p. 2.

Ahmed, S. (2006). *Queer phenomenology: Orientations, objects, others.* Durham and London: Duke University Press.

Alcoff, L.M. (2006). *Visible identities. race, gender, and the self.* New York: Oxford University Press.

Andreassen, R. (2007). *Der er et yndigt land: Medier, minoriteter og danskhed.* Copenhagen: Tiderne Skifter.

Andreassen, R. and Lettinga, D. (2012). Veiled debates: Gender and gender equality in European national narratives. In: S. Rosenberger and B. Sauer, eds., *Politics, religion and gender.* New York: Routledge.

Arendt, H. (1958). *The human condition.* Chicago: Chicago University Press.

Bech, H. and Necef, M.Ü. (2012). *Er danskerne racister: Indvandrerforskningens problemer.* Fredriksberg: Frydenlund.

Borren, M. (2008). Towards an Arendtian politics of in/visibility: On stateless refugees and undocumented aliens. *Ethical Perspectives: Journal of the European Ethics Network*, 15(2), pp. 213–237.

Brighenti, A. (2007). Visibility: A category for the social sciences. *Current Sociology*, 55(3), pp. 323–342.

Brown, W. (2006). *Regulating aversion: Tolerance in the age of identity and empire.* Princeton, NJ: Princeton University Press.

Cekic, Ö.S. (2014). Ja, nu er der plads til muslimer. *BT*, 20 June. Available at: http://www.infomedia.dk

Dyer, R. (1997). *White: Essays on race and culture.* London: Routledge.

Dæmpet velkomst (2014). *Jyllandsposten.* June 19. Available at: http://www.infomedia.dk

Fanon, F. (1967). *Black skin white mask.* New York: Grow Press.

Gale, R. (2005). Representing the city: Mosques and the planning process in Birmingham. *Journal of Ethnic and Migration Studies*, 31(6), pp. 1161–1179.

Goldberg, T. (2009). *The threat of race: Reflections on racial neoliberalism.* Oxford: Wiley-Blackwell.

Göle, N. (2011). The public visibility of Islam and European politics of resentment: The minarets-mosques debate. *Philosophy and Social Criticism*, 37(4), pp. 383–392.

Graham, S. (2011). *Cities under siege: The new military urbanism.* London: Verso.

Gregory, D. (2011). The everywhere war. *The Geographical Journal*, 177(3), pp. 238–250.

Hadj-Abdou, L., Rosenberger, S., Saharso, S. and Siim, B. (2012). The limits of populism: Accommodative headscarf policies in Austria, Denmark, and the Netherlands. In: S. Rosenberger and B. Sauer, eds., *Politics, religion and gender: Framing and regulating the veil.* Milton Park: Routledge.

Hervik, P. (2002). *Mediernes Muslimer: En antropologisk undersøgelse af mediernes dækning af religioner i Danmark.* København: Nævnet for etnisk ligestilling.

Hervik, P. (2008). Original Spin and its side effects: Freedom of Speech as Danish news management. In: E. Eide, R. Kunelius, and A. Philips, eds., *Transnational media events: The Mohammed cartoons and the imagined clash of civilizations.* Göteborg: Nordicom.

Hervik, P., Eide, E. and Kunelius, R. (2008). A long and messy event. In: E. Eide, R. Kunelius, and A. Philips, eds., *Transnational media events: The Mohammed Cartoons and the imagined clash of civilizations.* Göteborg: Nordicom.

Honneth, A. (1995). *The struggle for recognition: The moral grammar of social conflicts.* Cambridge: Polity Press.

Hussein, M. (2000). Media and minorities in Denmark. *Current Sociology*, 48(4), 95–116.

Hvorslev, B. (2014). Moskeer er uundgåelige i et demokrati. *Jyllandsposten*, June 11. Available at: http://www.infomedia.dk

Johansen, M. (2014). Spidserne bliver væk fra indvielse af stormoské. *Jyllandsposten*, June 15. Available at: http://www.infomedia.dk

Journal of Ethnic and Migration Studies (2005). Special issue: 'Mosque conflicts in European cities', 31(6).

Juul, K. (2014). Performing belonging, celebrating invisibility? The role of festivities among migrants of Serbian origin in Denmark and in Serbia. *Nordic Journal of Migration Research*, 4(4), pp. 184–191.

Kadher, N. (2014). Broderskabets moské, måske ikke. *BT*, June 21. Available at: http://www.infomedia.dk

Koefoed, L. and Simonsen, K. (2007). The price of goodness: Everyday nationalist narratives in Denmark. *Antipode*, 39(2), pp. 310–330.

Koefoed, L. and Simonsen, K. (2010). *"Den fremmede", byen og nationen – om livet som dansk minoritet.* Frederiksberg: Roskilde Universitetsforlag.

Leinonen, J. and Toivanen, M. (2014). Researching in/visibility in the Nordic context: Theoretical and empirical views. *Nordic Journal of Migration Research*, 4(4), pp. 161–167.

Lentin, A. and Titley, G. (2011). *The crisis of Multiculturalism: Racism in a neoliberal age.* London: Zed Books.

Loftsdottir, K. and Jensen, L. (2012). *Whiteness and post-colonialism in the Nordic region: Exceptionalism, migrant others and national identities.* Farnham: Ashgate.

Lægaard, S. (2010). Religiøse symboler, religionsfrihed og det offentlige rum: 'stormoskeer' i København. *Tidsskriftet Politik*, 13(4), pp. 6–14.

Nordic Journal of Migration Research (2014). Special issue: 'Researching in/visibility in the Nordic context', 4(4).

Rabo, A. and Andreassen, R. (2014). The Nordic discomfort with 'race'. *Nordic Journal of Migration Studies*, 4(1), pp. 40–44.

Shaw, W.S. (2007). *Cities of whiteness.* Hoboken: Wiley-Blackwell.

Simonsen, K. (2007). Practice, spatiality and embodied emotions: An outline of a geography of practice. *Human Affairs*, 17, pp. 168–182.

Simonsen, K. (2015). Encountering Racism in the (Post-)welfare state: Danish experiences. *Geografiska Annaler B*, 97(3), pp. 213–223.

Stormoské. (2014). *Politiken*, 20 June. Available at: http://www.infomedia.dk

Tebble, A.J. (2006). Exclusion for democracy. *Political Theory*, 34(4), pp. 463–487.

Thomsen, M. (2000). Luder og Madonna igen. *Information*, 14 June.

Yilmaz, F. (2016). *How the workers became Muslims: Immigration, culture, and hegemonic transformation in Europe.* Ann Arbor, MI: University of Michigan Press.

9 Mediating (in)visibility and publicity in an African church in Ghent

Religious place-making and solidarity in the European city

Luce Beeckmans

Blind spots: new patterns of African migration and new religious landscapes in Belgium

Over the last couple of decades, the number of sub-Saharan African migrants in Belgium has risen rapidly. Nevertheless, their presence seems almost invisible, both in the literature on urban migration, where we can speak about a 'a blind spot', but also in Belgian cities, where most Africans occupy highly interiorised spaces, often in peripheral locations. While some literature exists on the spatial implications of more settled immigrant groups in Belgium, for instance on religious facilities of Turkish and North-African population groups (Kanmaz, 2009), sub-Saharan African spatiality is an almost neglected topic of research. An exception is the 'Congolese' neighbourhood of Matonge in Brussels, directly tied to Belgium's colonial past (Arnaut, 2006; Demart, 2013). Changing migration flows and patterns of settlement over the past decades make the prevailing focus on Congolese diaspora in the Brussels metropolitan area to some extent obsolete, however. Since the turn of the century, more complex patterns of migration to Europe, including Belgium, have emerged as a consequence of the interplay of economic globalisation and trans-national network formation. With regard to migration from sub-Saharan Africa, this has resulted in the influx of migrants from more African countries of origin, as well as in a broader geographical distribution of African migrants over the Belgian territory, channelling African migrants not only to the capital Brussels but also to secondary cities or 'mid-sized cities', a geographic scale which is still largely unexplored in relation to migration (Glick Schiller and Çağlar, 2009).

The focus of this chapter is on some of the everyday spaces of Sub-Saharan African migration in Ghent, a mid-sized city in the Flemish part of Belgium. In particular, the chapter will take a close look at the processes and products of physical place-making by African migrants 'from below', i.e. self-organised and outside any institutionalised framework of intervention. However, while I will attempt to highlight the agency of African migrants in contemporary processes of urban place-making, I will try not to fall into the 'post-colonial trap' of overestimating or romanticising the power of such interventions. Indeed, as this chapter will demonstrate, African presence and place-making in the city does not necessarily translate into political presence or actor-ness, but involves, rather, a complex and ongoing negotiation over degrees of (in)visibility and levels of publicity. By focussing on instances of African place-making, this chapter is in line with recent advances in both migration studies and urban studies, in which migrants are increasingly considered as constitutive forces in the shaping and making of contemporary cities in Europe (Çağlar and Glick Schiller, 2018). At the same time, it hopes to add to such studies that have focussed on the role of migrants as 'scale makers' in the global 'scalar' repositioning of cities in more abstract fashion (Glick Schiller and Çağlar, 2011), or on the remaking of the social spaces of cities through the encounter of multiple 'everyday diversities' or 'lived diversities' (Van Dijk, 2011; Wessendorf, 2014; Husband et al., 2014). The question of how physical space itself shapes the unfolding of diversity and, conversely, how physical space is being shaped by diversity are dimensions that have been rather under-developed until now (Berg and Sigona, 2013, p. 356; Erdentug and Colombijn, 2007). Only a few (co-edited) works such as *Drifting. Architecture and Migrancy* (Cairns, 2004) and *Ethno-architecture and the politics of migration* (Lozanovska, 2015) explicitly dwell on the materiality of immigrant space. My contribution aims to complement these studies.

In order to expand our understanding of the materiality of sub-Saharan immigrant space, I will discuss here one manifestation of African spatiality in Ghent more deeply: an African church located in an ordinary terraced house and, therefore, quite invisible within the urban fabric. In Ghent, as elsewhere in Europe, the arrival of sub-Saharan African migrants has resulted in the rise of Afro-Christian churches, more specifically Pentecostal churches, which are said to deploy a 'reverse mission' to Europe when compared to Catholic missions to Africa in the colonial era (Robbins, 2004; Freston, 2010). By investigating into the spatial dimension of Afro-Christian religiosity, this chapter furthermore connects with a 'spatial turn' in research on religion that has begun to explore the interrelationships between religion, space, and place (Knott, 2010; Hervieu-Léger, 2002), and in which African Pentecostalism takes a prominent role within the wider context of global religions (Becker et al., 2014; Adogame and Shankar, 2013). In other work, I have analysed African religious place-making, first, by foregrounding new geographies of Afro-Christian (spatial) conversion, in which mid-sized cities take a prominent role (Beeckmans, 2017); second by conceptualising the spatiality of African churches as alternative, and often precarious, nodes of urban regeneration (Beeckmans, 2020a). In this chapter, I will focus mainly on the dialectics between visibility and invisibility in relation to African religious spatiality and public space.[1] My excavation in Ghent's post-industrial urban tissue thus is an endeavour in "presencing absence", "foregrounding the subaltern, life on the margins, the ordinary" (Harrison and Schofield, 2010, p. 10), "materializing that which is forgotten or concealed" (Buchli et al., 2001, p. 171), and, as such, is inspired by some recent scholarship in the field of archaeology of the contemporary past, and its focus on material culture (Holtorf and Piccini, 2009). By critically shedding light on the spatial dimension of current processes of trans-national migration, and the related questions of (geopolitical) power, as well as processes of spatial inclusion and exclusion, this chapter also engages broader questions of a 'politics of space' (Certomà et al., 2012; see also Gonzalez-Ruibal, 2008) and, especially, of the negotiation of lived difference in space (much like Swanton's chapter in the opening section). All the while, by focussing on 'real' spaces and their 'overlooked' materialities, I also attempt to make some of the more abstract, post-colonial concepts, such as 'hybridity' or 'appropriation' more concrete and tangible (Teverson and Upstone, 2011), in particular by highlighting their spatial dimension.

The chapter draws on ethnographic fieldwork I have conducted as a post-doctoral research fellow during the period 2016–2019 in several locations in Ghent, which is both my place of residence and my workplace, as well as on the fieldwork of some of my students.[2] The resulting 'ethnography of space' (Low, 2017) is based on semi-structured interviews with some key informants, as well as participant observation in the church. In addition, I have mobilised my skills as an architect to visualise the ethnographic fieldwork in drawings, used both as an instrument of analysis (fieldwork drawings) and as a way to convey the argument to the reader (illustrations). This way of working is close to Hall, King, and Finlay (2015, p. 59) approach, which also combines ethnographic and architectural methods, using drawing as an "exploratory and critical visual practice, providing us tools to see socio-spatial relationships in temporal and scalar dimensions". In contrast to photographic material, the abstractness of the drawing is also mobilised as a way to protect the privacy of the informants, as well as to anonymise the intimacy of their places. This chapter contains one figure (prepared in collaboration with a professional illustrator), whose aim is not to offer a one-to-one projection of reality, but instead to envision a highly interiorised, and as a consequence quite invisible, African space in Ghent in a textured way.

African religious spatiality in Ghent: mediating (in)visibility and publicity

In order to conceptually frame the empirical analysis of the African church in the following section, I will first expand upon some theoretical aspects regarding the (in)visibility of such 'public space'. I place the term 'public' in quotation marks because even after years of intense fieldwork it remains an open question to me to what extent this place is actually 'public'. My main informant, the pastor of the church, tends to describe the place as 'public', pointing to the fact that no formal membership is necessary to access the church. Indeed, already on my first fieldwork visit, I was welcomed with great hospitality that is quite uncommon in most public urban spaces in Europe. At the same time, while the church ostensibly enables encounters between strangers from different backgrounds, most users are of predominantly African descent. Yet, as I will show here, the degree of publicity varies throughout the building, as well as through time, as domestic and more public facilities are entangled in peculiar ways in the terraced house. Altogether, I would rather describe the African church as 'parochial', as relations between people are less ephemeral and contingent than in a public place, but instead more marked by commonality and conviviality (Wessendorf, 2014; Werbner, 2002). In that sense, the church might be captured by what Amin (2002) describes as 'micro-publics' in his seminal article "Ethnicity and the multicultural city: living with diversity". These are places where 'prosaic negotiations' between people are compulsory and therefore can often lead to 'meaningful encounters' (Valentine, 2008, p. 325; Fincher and Iveson, 2008; Merrifield, 2012; see also Swanton's chapter in this volume). Yet, as will become evident from the socio-spatial analysis in the next section, the relationship between visibility and publicity is not always as straightforward as is often the case in other micro-public spaces (Staeheli et al., 2009).

The dialectics of visibility and invisibility in relation to African religious spatiality and public space has been the object of various other scholarly analyses. David Garbin (2013, p. 682) for instance argues that the experience of place-making of African churches is often "synonymous with invisibility, and sometimes even with instability and precarious territoriality" since the spaces occupied by Africans are mostly located "within the gaps of the post-industrial city", often interstitial or marginalised urban sites such as former warehouses or depots (see also Dejean, 2010; Knowles, 2013; Baird, 2014). Similarly, and also in context of post-industrial London, Kristine Krause (2008, pp. 109, 110) discusses how Africans have established places of worship "hidden behind scrap metal and next to dealers' garages and repair shops" by transforming and appropriating non-religious buildings while at the same time attempting to "[meet] certain juridical, political and financial requirements". Others, such as recently Ayse Çağlar and Nina Glick Schiller (2018, p. 155), have discussed African churches in relation to their 'social citizenship claims' arguing that "based on what they see as their God-given rights, migrants moving within a transnational social field endeavor to act upon the institutional, legal and societal processes of the state and locality in which they have settled to claim place-based rights and identities". What is clear, and what will be further developed in the empirical analysis, is that African churches are much more than religious sites, but in quite 'invisible' ways also fulfil essential political, social, cultural, psychological, and economic needs, which have not been accounted for by the state (in spatial terms) for this group, thus forming alternative networks of solidarity. As such, they "provide outreach beyond the prescriptions of faith or persuasion" and sometimes "[fill] in the gaps in care left by a receding state" (Hall et al., 2016, p. 9).

Whereas African negotiations over visibility/invisibility in the public realm still remain somewhat understudied, this is very much in contrast to the abundance of studies on Muslim (in)visibility in European cities. While Islam is today hypervisible in the academic and public debate, Muslim place-making in Europe is also to a large extent characterised by invisibility (Eade, 2011; Gale and Naylor, 2002; Oosterbaan, 2014). Two reasons behind this invisibility are in particular insightful to dwell a little longer on, as they echo to a certain extent my research findings on African spatiality in Ghent. First, some authors, such as Benchelabi (1998) in her study of a 'hidden' network of *salons de thé* for Maghrebi women in Brussels, question the universality of what is commonly considered as 'public' and 'private' in Belgium. Benchelabi (1998, p. 5) for instance highlights a certain inversion of these two notions (at least within a Eurocentric vision), and relates this to a gender-specific division of space within Muslim communities: "In Brussels, the domestic space is public and private at the same time: It organizes all forms of social, religious, and family gatherings. It takes on new meanings. It is a form of resistance of an alienating 'outside'". Similar to the two African spaces discussed later, on a material level, this results in a high interiorisation of (public) space by means of a recurring set of elements such as opaque curtains and draperies. Exactly because Muslim women, and for other and similar reasons African people, regularly feel a certain disinterest and discomfort in European public space, the 'space of appearance', with its specific, but often unfamiliar, visibility regimes and hegemonic ideals of 'the' public sphere (Staeheli, Mitchell and Nagel, 2009, p. 633); they turn, also for (semi-)public purposes, to the intimate and "'private' sphere of 'invisibility', a space of 'mere giveness' that allows the (political) subject, the citizen, to be just what she or he 'is'" (Arendt, 1943; cited by Bialasiewicz, 2017, p. 382).

A second parallel between African and Muslim spatialities in Europe in relation to the visibility/invisibility dialectics (even though most Africans I have met in Ghent are Christians) is that many African migrants, often deliberately, opt to stay under the radar for legal reasons. Apart from the fact that many of the Africans I encountered in the church are irregular migrants who do not hold formal status, also some of their place-making practices conflict directly with the regulatory regimes of the Belgian state, for instance municipal laws on subletting, laws on the consumption of alcohol, building and fire safety regulations, etc. Gaining visibility in urban space is also linked to notions of belonging and identity and, in the context of Muslim architecture, some authors, such as Bialasiewicz (2017, p. 381) writing about the negotiation of Islamic spaces of worship in Italy, have noted how what appear to be Muslim claims on public space are often met with feelings of discomfort and even fear, resulting in strategies of invisibility: "as long as it [the mosque] is an inconspicuous structure [. . .] 'no one will complain'". Indeed, Bialasiewicz (2017, p. 377) also describes authorities' attempts to shut down (also 'authorised') mosques by appealing to 'public health reasons' or sanitary and fire safety regulations as a technical pretext (see also Beynon, 2015; Frazier, 2015). During my fieldwork in Ghent I have encountered similar negotiations over visibility and invisibility, both at the side of the African pastor, and the municipality, as will be explained later. In my analysis, I have paid particular attention to the specificity of African migrant agency/identity (related to aspects of ethnicity, gender, and religion) in relation to hegemonic regulatory regimes.

An invisible church in Ghent: mediating liminality and solidarity

In Ghent, most migrants from African descent originate from Ghana and Nigeria. Although still relatively small in size today, they are the fastest growing immigrant group in Ghent after East Europeans (Stad Gent, 2017). In Ghent there exists a concentration of African spatiality on the eastern periphery of the city, around the railway station and along the former docks. In this area the African church discussed in this chapter was first located on two different sites, but since a couple of years the church has moved to another location on the complete other side of the city. Yet, while there exists a certain African concentration on the east-side of Ghent, it would be far from appropriate to speak about an African 'ethnopolis' as it is the case for Matonge in Brussels, as Africans still form a minority in this neighbourhood and are also dispersed over other neighbourhoods of the city. We could rather say that this area, just like some other parts of the 19th-century belt of the city in between the inner and outer ring road, forms a location with a certain 'establishment potential' (Vyzoviti, 2005) for migrants from sub-Saharan Africa due to a large number of vacant properties, and thus affordable rents, as well as its high accessibility for public transport. In this area, Africans have, through the occupation and appropriation of ordinary houses, warehouses, shops, garages, railway infrastructure, and temporary structures, transformed existing non-religious buildings into churches (some African churches in Ghent are integrated in existing church-buildings, but these churches are not established by Africans and are not the topic of this chapter). As such they have territorialised under-regulated and underutilised spaces, left vacant in the post-industrial city, because they lack functionality to the immediate local setting.

By being located on the west-side of the city, the church discussed here, a parish of the Redeemed Christian Church of God (RCCG), one of the largest Pentecostal churches in Europe with its roots in Nigeria, has a rather isolated position. Its pastor Kofi (all names are anonymized), who came to Belgium like many of his colleagues via a student visa while having a clear religious mission since he was already pastor in Ghana, nonetheless identified several advantages to the site. When I asked him why he deemed this place suitable to establish a church, he immediately pointed to a bus stop in close proximity, as well as to the availability of free parking spaces in front of the house, something which was largely lacking at the previous locations of the church in Ghent which resulted in many traffic violations, as well as complaints from neighbours and the police fanatically writing out fines during the services. The main disadvantage of the location, according to Kofi, is its remoteness from a railway station, as many worshippers also come from other cities in Belgium. In order to solve this problem, Kofi transformed his car into a taxi service, with the logo of the church on the side doors, and before each service he drives to several locations to pick up his worshippers. This way he is able to fulfil one of RCCG's most important points of its mission statement, available on and widely distributed through the Internet, namely the duty to "plant churches within five minutes walking distance in every city and town of developing countries and within five minutes driving distance in every city and town of developed countries". However, while the RCCG has a very pronounced territorial mission, several conflicts over territory have occurred within the church itself, and Kofi's move to the west-side of the city can also to be viewed as a result of this. One of the territorial strategies of the RCCG is the requirement of each 'mother church' to plant four satellite churches (Beeckmans, 2017). Yet, this process is not always as controlled as the 'general overseer' in Nigeria, and above all the 'assistant general overseer' who is responsible for Europe Mainland, as well as the Belgian head of the RCCG, who I both interviewed respectively in Amsterdam and Antwerp, would like to see it. Indeed, while Kofi's church was established by a 'mother church' in Nigeria, another RCCG church was simultaneously being created in Ghent according to a more official policy and to avoid an overlap in territory and a competition over worshippers, Kofi decided to migrate to the other side of the city.

However, while the RCCG is on an explicit mission to expand its activities across Europe by emplacing churches in more or less coordinated distances from each other, and in this way no longer just covering capital or global cities, this process unfolds quite invisibly. In Belgium, the presence of Afro-Christian churches is, unlike mosques, only very seldom part of the public debate and, when addressed, is either depicted as an 'exotic' phenomenon due to its 'sparkling liturgy' or considered problematic because of its uncommon rituals.[3] For instance, about ten years ago, Pentecostal churches exceptionally made the news, when the Federal Information and Advisory Centre for Harmful Sectarian Organisations ('Informatie en Adviescentrum inzake Schadelijke Sektarische Organisaties') expressed its 'deep concern' about the Pentecostal movement in Belgium, because the churches would be harmful to African immigrants "since many worshippers were forced to stop medical treatments and instead compelled to turn to exorcism" (in my experience they rather seem to combine different treatments simultaneously).[4] The expectation of worshippers to hand over one tenth of their income for the functioning of the church as an 'offering' also captured the eye of the press at a certain point, but the news soon vanished as the public lost interest.[5] Hence, while the presence of Afro-Christian Churches is exponentially rising, with now already more than 20 Pentecostal churches in Ghent only, this is for predominantly an invisible kind of presence. This invisibility in the public arena is partly a consequence of the material invisibility of the churches. Indeed, when looking at Kofi's church, only two modest banners on the façade, one black and white and one in with the yellow and green coloured logo of the church, reveal its function as a church, which is even to RCCG standards quite invisible. The main reason for this is that Kofi believes that by keeping his church, registered as a non-profit organisation since it is not officially recognised (hence not receiving state funding), 'camouflaged' in the streetscape as an ordinary terraced house, he will be able to avoid problems with neighbours. Instead of capturing attention, he prefers to blend into the streetscape. Additionally, he explains that most worshippers find their way to the church via the Internet, making strong visible markers on the façade unnecessary to attract worshippers. As such, the invisibility of RCCG churches in European streetscapes seems to go hand-in-hand with a high visibility and activity on the web (for instance the church locator on the RCCG Europe website or Kofi's church's own YouTube channel and Facebook page), which is the main access point for the church worshippers but also for connection with other worshippers worldwide. This salient presence on social media and the

Internet well illustrates the trans-national life of many Africans, who both online and offline are highly mobile and interconnected (Beeckmans, 2019; Fortunati, Pertierra and Vincent, 2012).

Kofi and his wife Olivia are currently renting the small house from a man originally coming from Aruba, but are now living in the Netherlands. What was particularly attractive to Kofi when he visited the building for the first time was the large, free-standing depot in the back of the house, at the time accommodating a fitness centre. In the past, when his church was still housed in a terraced house in the area around the eastern railway station, Kofi encountered many difficulties with neighbours complaining about noise pollution, with regular police interventions as a result. As Kofi's services, on Sunday mornings, and sometimes also on Tuesday and Thursday evenings, are characterised by loud, live music, group singing and dancing, the depot has the advantage of lacking close neighbours, and thus being what he calls 'soundproof'. Also from other (RCCG) pastors I learnt that the requirement to be 'soundproof' is one of their top priorities when looking for a location. It is also the main reason behind many of their frequent relocations.

There is another reason for the invisibility of the church. Indeed, while the façade only and in very modest ways marks the church's function, the ordinary terrace house in fact accommodates much more than simply a church. Instead it rather functions as a 'public space' or a new collective world outside the urban regularity. As will be shown in the following section, apart from being a place of worship, the house also functions as a place for gathering and meeting, for sharing food and drinks, for counselling services and for care-taking and accommodation for worshippers that frequent the church, as well as for new-incoming migrants. As such, almost as important as the religious services are the social services provided by the church. Both have a profound materialisation in the ordinary terraced house.

When we enter the house via a small front door, we immediately, without any threshold, enter the large kitchen of the family, which also serves as the main meeting room for the worshippers besides the worship hall. Throughout the house it is remarkable how much the church (related) spaces and the family's domestic spaces are completely intermingled. The kitchen only has a small window, but since there is a thick, grey curtain in front of it, day light and undesired glimpses are kept out and a more intimate sphere is created. Everywhere where possible, sofas, tables, and chairs are placed, in total providing a seating capacity for around 30 persons. On the side are two large freezers. After every Sunday service, on an average taking three hours, worshippers share a meal, prepared by the women at home beforehand or in the kitchen of the church house. For some worshipers it forms a chance to have an extra meal, as many have to cope with financial problems. Many authors have pointed to the key importance of food and food culture within the diaspora, as food and food-related practices trigger cultural memories and enable connection with the wider diasporic group through processes of belonging and identity-formation, albeit never without certain degrees of adaptation and hybridisation (Mintz, 2008; Sen, 2015, 2016; Watson and Cardwell, 2005). Yet African food culture also has a profound impact on place-making as the

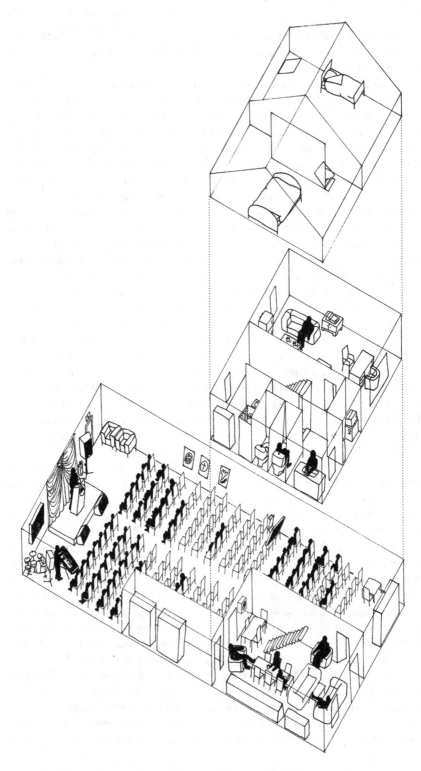

Figure 9.1 Sketch of the Redeemed Christian Church of God, Ghent

integration of a canteen-like space, covering the entire ground floor of the house and taking over the family's dining room, leads to a profound cultural and physical transformation of what could be a regular terraced house in Ghent. On weekday evenings, after pastor Kofi returns from his regular job, the kitchen is, apart from a place for bible study, also a meeting-place for worshippers. There people enjoy each other's company, but equally solve financial issues, for instance though a loan system, in case someone of the church community has an unexpected major expense to cover. Therefore, already after the first steps into the house, it becomes clear that the church is much more than simply a place of worship.

From the kitchen, there are two ways to enter the worship hall. In the back is a sliding door, but the main entrance, clearly signed by a hand-drawn signboard, is a door giving out to the garage on the right. While there is still a garage door visible on the façade, the place no longer functions as such and the outdoor connection is blocked with a large closet. The garage is now an extension of the worship hall, containing 30 seats with a television screen allowing people in the back to follow the service. When we walk a little further, we enter the surprisingly large worship hall, completely lighted with bright neon-lamps in the ceiling, as there is only a door with a tiny window at the other side of the hall providing direct sunlight. Another universe, a microcosm, is created here, symbolically connecting to other RCCG parishes in Europe as well as Africa. As elaborated in some earlier articles (Beeckmans, 2017, 2019), migrant place-making practices are made possible through both physical/material as well as virtual trans-urban networks, resulting in similar space compositions and decorations. Also here, according to pragmatic and even hyper-functional logic, the recurring set of elements consisting out of plastic chairs, draped curtains, red carpets, vast musical equipment, television screens, recording machinery (as the service is broadcasted on YouTube and followed from several locations in and outside Belgium), electric heaters (since the central heating system is out of office), and loud posters give form to a typical RCCG church.

At both sides of the small corridor, ten rows of five seats are placed. Although Kofi dreams of being able to hold two alternating services on Sunday, the sign of a successful church, on an average service, only half of the hall is filled with women, children, and men, around 70 persons in total. Yet, during the service, more people come in and by the end of the service, 75% of the seats are taken. On several occasions Kofi reassured me that also Belgian people come to his services, this way stressing his contribution to the 'reverse mission' to re-convert secularised Europeans, but during the bilingual (French and English) services that I attended, only a few 'whites' (Kofi's words) were present. Most people were of African descent, yet from all kind of nationalities, not only Nigerians. Therefore, until today, Kofi's mission has been only partly successful: while his church is now accommodating a vibrant community, his mission to convert natives is still unaccomplished. This is, however, something Kofi is really aiming for, as he thinks it will give his church both recognition and legitimacy. It is telling that a couple of years ago he enthusiastically cooperated with a news reportage, as he believed it was a way to attract more native worshippers.[6] He proudly talked about the reactions, also from Belgians, the reportage evoked in the following weeks, yet without the result he hoped for. Kofi's wife and French-speaking son in law, originating from Cameroun, frequently also take the lead in the services, and, as they are called upon by Kofi, the worshippers "dance as if it is the last day they can dance". Also highly visible in the hall are the many fire extinguishers and emergency lights. It is obvious that Kofi has made every effort to comply with fire regulations. Asked for the reason, Kofi said that at the former location, he had the feeling that the police often tried to expel him from the building by using violations of fire safety regulations as a pretext. Consequently, emergency lights, and extinguishers, as well as, at other locations, acoustic panels, are as much part of an Afro-Christian church in Europe as the altar is.

To a large degree, the church forms an incremental place. A place which is daily being built 'from below' and which is in an eternal unfinished state: in the dreams of its pastor, and on the Internet. An example of this is the model of a church building in Kofi's office. We are now on the first floor, coming here via a large stairway in the kitchen. The pastor's office is the most important room on this floor and also the most comfortable of the house. Here Kofi receives his individual guests for spiritual conversations but also for more worldly and practical consultations, resembling counselling sessions. Apart from his writing desk with computer and seats at both sides and a small freezer with soft drinks, this model captures the visitor's attention. Made by a Belgian architect active in the Christian movement, the model is the embodiment of Kofi's aspiration to build one day a large and modern church building in Ghent, giving himself and his church more prestige, as well as visibility. As a preferred setting, Kofi points to the area where also the new football stadium is currently located, as there he and his worshippers are unlikely to disturb other people with their music and signing. In order to take some steps forward with his project, Kofi already had a meeting with the former, socialist mayor of the city. Also in other cities, I noticed how pastors carefully tried to build up some political capital, also for instance by inviting politicians to their church. It is remarkable that by doing so they tend to stress their extra-religious services: for instance the fact that they take care of newcomers, distribute meals, organise day-care, prevent drug dealing and keep African women out of prostitution.[7] The ultimate aim of this lobbying is to create a certain "visibility [and accompanied legitimacy] on the religious landscape" in Belgium (Maskens, 2012, p. 399). Remarkably, the model of the church and the rendition of the building on YouTube look a little like a football stadium. While Kofi is frustrated that he has to hold his services in a former fitness centre while at the same time the services in Ghent's cathedrals attract almost no worshippers, it's certainly not the type of building he is dreaming about. Instead Kofi aims to establish a modern building for what he calls a 'modern religion', opposing it in this way to the old cathedrals of the Catholic Church in Belgium. Some authors even go as far as to speak about "entrepreneurial religion in the age of neoliberal urbanism" (Lanz and Oosterbaan, 2016, p. 487) and when we look at Kofi's growth scheme also his church's aspirations fit into this description.

On the floor where Kofi's office lies, we also find a bathroom with a separate ladies' and men's room. The toilets are marked with signs that we might see in any other public building, even though here they look unnecessary as the house is very tiny. Strangely, but out of necessity, these are also used as the private toilets of Kofi and his wife. On the other side, and also seemingly public, as one just walks in, is a shower and a washing machine. Yet it is probably the most private place of the house. A small floor leads us to the second floor, where under the roof two sleeping rooms are organised. The largest one is of Kofi and his wife, the second one was first used by the daughter, but as she now lives somewhere else, it now provides room for some occasional guests lacking a sleeping place in the city.

New (invisible) collective worlds in the post-colonial city in Europe

As highlighted in this chapter, African churches can be considered as one of the materialisations of the trans-national lives many migrants lead. Indeed, the church discussed in this chapter can be conceptualised as a node in a globe-spanning network that interconnects migrants worldwide (Beeckmans, 2017; Becker et al., 2014). Nevertheless, the church is also strongly bound to its current locality. Here, global relationality and territorial embeddedness are deeply intertwined. The more extensive (and also virtual) spaces of Afro-Christianity are made real here through distinct, physical modes of place-making, adapted to their specific context (see also the discussion in Beeckmans, 2019). The result are 'hybrid' architectural forms, also called 'mutualisms' by Hall (2015, p. 32). Although this hybrid architecture has been described by some as a 'migrant aesthetics' (Lozanovska, Levin and Gantala, 2013), marked by specific ethnic 'markers' and 'identifiers', or even 'third world–looking architecture' (Beynon, 2005) because it often remains in an eternal, unfinished state, waiting on new financial injections to come, Africans at the same time "engage, in their own terms, in a transformative project of spatial appropriation, regeneration and re-enchantment of the urban landscape" (Garbin, 2013, p. 677), sometimes contributing to alternative ways of urban regeneration (Beeckmans, 2020b; Çağlar and Glick Schiller, 2018). Therefore, even if African migrancy does not always produce new sites of insurgent citizenship and often forms spaces of marginality, African place-making such as that noticeable at the RCCG church in Ghent can open up spaces where the 'right to the city' (Lefebvre, 1968) can be, at least partially, claimed and re-negotiated. Indeed, through Afro-Christian religious place-making, new collective worlds outside the urban regularity are created in European cities, forming alternative spaces of support and solidarity among African migrants and acting as an 'arrival infrastructure' for newcomers (Meeus, Arnaut and van Heur, 2019). By organising parallel networks of solidarity and forming important nodes of self-organisation (Schillebeeckx, Oosterlynck and De Decker, 2019), these churches partially fulfil the role of the local state, which may explain the reason why they are 'tolerated' and sometimes even deliberately remain unseen by the local authorities.

This brings me, in closing, again to the question of visibility and its relation to the question of tolerance (of people, practices – and architectural forms) in today's European cities. From the empirical study of the church, it became obvious that the process of materialisation of trans-national lifestyles and connections, for instance by semi-fixed, stylistic appropriations of the façades (logos, signboards, banners, etc.) and above all significant architectural interventions in the buildings' interiors, is always a balancing act between the necessary visibility to function as a (semi-)public place and the required invisibility to circumvent hegemonic regulatory regimes. If we use the conceptual framework of De Certeau (1984), the place-making practices underscoring these materialisations can sometimes be understood as deliberate 'strategies' to manoeuvre within a legislative context marked by a deeply Eurocentric, normative and secularised conceptualisation of urban/public space. Yet, just as much, African place-making is simply the crystallisation of the manifold everyday 'tactics' African migrants deploy to make a living, without ever having the intention to break with the prevailing legislation or resist against dominant regimes of publicity. As has been argued by Casas Cortes et al. (2014), although in the context of border policies, "Most of the time, migrant struggles are concerned with neither representation nor claims for rights nor border policies as such. Rather, they are struggles of (migrant) everyday life: they consist in the mere fact of persisting in a certain space, irrespective of law, rights and the pace of the politics of mobility". Some scholars, such as Knowles (2013, p. 651) have declared that with their research they aim to resolve the invisibility of African spatiality in Europe's urban space as they consider "visibility vital in generating more open forms of urban encounter and, ultimately, citizenship". However, while this seems a legitimate intention, many of the place-making practices of African migrants just seem to require a certain degree of invisibility (see also Staeheli, Mitchell and Nagel, 2009, p. 635). Hence, making the coping mechanisms of migrants visible on the (political) public stage may also be dangerous, as this could result in an even higher fragility and instability of these groups. This is a difficult dilemma, which this chapter will not resolve, and that requires a profound reflection on the position of the researcher and the ethics of research. If I wanted to make something visible with this chapter, however, then it was, above all, the manifest paradox between "the role of the material world in shaping ethnic identity" and "the strategy of invisibility" deployed by African migrants in this African church in Ghent (Frazier, 2015, p. 48).

Notes

1 Elsewhere, I have analysed an African shop in Ghent from a similar perspective (Beeckmans, 2020b).
2 My FWO post-doctoral research is entitled "Mapping the Invisible City. Spatial Manifestations of sub-Saharan African Diaspora in the mid-size city in Europe (the case of Belgium)". I am in particular grateful to the following students, whose fieldwork within the framework of their MA research was insightful for this article: Pieterjan Dehaene (2014) *Ruimtelijke Manifestaties Van Afrikaanse Diaspora En Religie In De Middelgrote Stad: Een Typologisch Onderzoek Naar De Afrikaanse Kerken In Het Stedelijk Weefsel*

Van Gent as well as Zuzanna Rucka and Helena Sileghem (2017) *The Reversed Mission Between Discourse and Practice: Exploring Transnational Place-making Strategies of Afro-Pentecostal Churches In the Mid-sized City In Belgium (Aalst and Oostende)*.

3 Quote from "Afrikaanse Belgen, een weinig gekende groep op zoek naar zijn plek", Erik De Smet & Jozefien Van Huffel, Kerk & Leven, 21 February 2018.

4 "Adviescentrum sekten bezorgd over Pinksterbeweging", *Het Laatste Nieuws*, juni 2008. See also, "Pak de pinksterpastoors aan", *Mo Magazine*, Chika Unigwe, 22 November 2013.

5 "Evangelische en Pinksterkerken. Reportage", *De Standaard*, 25 March 2005.

6 Reportage of VRT NEWS, the prime television channel in Flanders. Source anonymised for privacy concerns.

7 These points were made both in my interview with the 'assistant general overseer' who is responsible for the European Mainland, as well as the Belgian head of the RCCG, whom I interviewed respectively in Amsterdam and Antwerp in 2017.

References

Adogame, A. and Shankar, S., eds. (2013). *Religion on the move! New dynamics of religious expansion in a globalizing world*. Leiden: Brill Publishers.

Amin, A. (2002). Ethnicity and the multicultural city. *Environment and Planning A*, 34(6), pp. 959–980.

Arendt, H. (1943). We refugees. Reprinted in *Altogether elsewhere: Writers on exile* (M. Robinson, ed.). London: Faber & Faber, pp. 110–119.

Arnaut, K. (2006). Blowing bubbles in the city or does urban governance have a bad breath? Reports from Matonge, somewhere in Brussels. *Aprior*, Dec., pp. 58–69.

Baird, T.E. (2014). The more you look the less you see: Visibility and invisibility of Sudanese Migrants in Athens, Greece. *Nordic Journal of Migration Research*, 4(1), pp. 3–10.

Becker, J., Klingan, K., Lanz, S. and Wildner, K. (2014). *Global prayers: Contemporary manifestations of the religious in the city*. Zürich: Lars Müller Publishers.

Beeckmans, L. (2017). Mobile Urbanism from below: Exploring afro-Christian churches as scale-makers and place-makers in European mid-sized cities. Paper presented to the Annual Conference of the Royal Geographical Society.

Beeckmans, L. (2019). Migrants, mobile worlding and city-making: Exploring the trans-urban circulation and interconnectedness of diasporic world-making practices. *African Diaspora*. Forthcoming.

Beeckmans, L. (2020a). Afro-Christian churches as (invisible) care-takers in/of the city: Between precarious occupation and dynamic appropriation of the built environment. In: F. Ferdous and B. Bell, eds., *All-inclusive engagement: Architecture in the age of global Austerity*. London: Routledge.

Beeckmans, L. (2020b). Postcolonial manifestations of African spatiality in Europe: The invisible 'public' spaces of Ghent. In: M. Lorenzon and J. Nitschke, eds., *Analyzing architecture and the built environment in the postcolonial era*. Berlin: Springer (Archaeology and Decolonization Series).

Benchelabi, H. (1998). Cultural displacement in Brussels with Maghrebi women. *Journal of Architectural Education*, 52(1), pp. 3–10.

Berg, M.L. and Sigona, N. (2013). Ethnography, diversity and urban space. *Identities*, 20(4), pp. 347–360.

Beynon, D. (2005). Melbourne's 'third world-looking' architecture. In: *Suburban fantasies: Melbourne unmasked*. Melbourne: Australian Scholarly Publishing, pp. 68–83.

Beynon, D. (2015). Edge of centre: Australian cities and the public architecture of recent immigrant communities. In: M. Lozanovska, ed., *Ethno-architecture and the politics of migration*. London: Routledge.

Bialasiewicz, L. (2017). 'That which is not a mosque': Disturbing place at the 2015 Venice Biennale. *City*, 21(3–4), pp. 367–387.

Buchli, V., Lucas, G. and Cox, M. (2001). *Archaeologies of the contemporary past*. London: Routledge.

Cairns, S., ed. (2004). *Drifting: Architecture and migrancy*. London: Routledge.

Çağlar, A. and N. Glick Schiller. (2018). *Migrants and city-making: Dispossession, displacement, and urban regeneration*. London: Routledge.

Casas-Cortes, M. et al. (2014). New keywords: Migration and borders. *Cultural Studies*, 29(1), pp. 55–87.

Certomà, C., Clewer, N. and Elsey, D. (2012). *The politics of space and place*. Cambridge: Cambridge Scholars Publishing.

De Certeau, M. (1984). *The practice of everyday life*. Berkeley, CA: University of California Press.

Dehaene, P. (2014). *Ruimtelijke Manifestaties Van Afrikaanse Diaspora En Religie In De Middelgrote Stad: Een Typologisch Onderzoek Naar De Afrikaanse Kerken In Het Stedelijk Weefsel Van Gent*. Unpublished MA thesis, University of Ghent. Available at: https://lib.ugent.be/fulltxt/RUG01/002/153/513/RUG01-002153513_2014_0001_AC.pdf

Dejean, F. (2010). La précarité spatiale des églises africaines en Seine-Saint-Denis et sur l'Ile de Montréal. In: S. Fancello, and A. Mary, eds., *Chrétiens Africains en Europe: Prophétismes, Pentecôtismes et Politiques des Nations*. Paris: Karthala.

Demart, S. (2013). Congolese migration to Belgium and postcolonial perspectives. *African Diaspora*, 6, pp. 1–20.

Eade, J. (2011). Sacralising space in a Western, secular city: Accommodating Muslim and Catholic migrants in London. *Journal of Town and City Management*, 1(4), pp. 355–363.

Erdentug, A. and Colombijn, F., eds. (2007). *Urban ethnic encounters: The spatial consequences*. London: Routledge.

Fincher, R. and Iveson, K. (2008). *Planning and diversity in the city*. London: Palgrave Macmillan.

Fortunati, L., Pertierra, R. and Vincent, J. (2012). *Migration, diaspora and information technology in global societies*. New York: Routledge.

Frazier, J.W. (2015). Indian-American landscapes in Queens, New York. Ethnic tensions in place-remaking. In: M. Lozanovska, ed., *Ethno-architecture and the politics of migration*. London: Routledge.

Freston, P. (2010). Reverse mission: A discourse in search of reality. *PentecoStudies: An Interdisciplinary Journal for Research on the Pentecostal and Charismatic Movements*, 9.

Gale, R. and Naylor, S. (2002). Religion, planning and the city: The spatial politics of ethnic minority expression in British cities and towns. *Ethnicities*, 2(3), pp. 387–409.

Garbin, D. (2013). The visibility and invisibility of migrant faith in the city. *Journal of Ethnic and Migration Studies*, 39(5), pp. 677–696.

Glick Schiller, N. and Çağlar, A. (2009). Towards a comparative theory of locality in migration studies: Migrant incorporation and city scale. *Journal of Ethnic and Migration Studies*, 35(2), pp. 177–202.

Glick Schiller, N. and Çağlar, A. (2011). *Locating migration: Rescaling cities and migrants*. Ithaca, NY: Cornell University Press.

González-Ruibal, A. (2008). Time to destroy: An archaeology of supermodernity. *Current Anthropology*, 49(2), pp. 247–279.

Hall, S.M. (2015). Super-diverse street: A 'trans-ethnography' across migrant localities. *Ethnic and Racial Studies*, 38(1), pp. 22–37.

Hall, S.M., King, J. and Finlay, R. (2015). Envisioning migration: Drawing the infrastructure of Stapleton Road, Bristol. *New Diversities*, 17(2), pp. 59–72.

Hall, S., King, J. and Finlay, R. (2016). Migrant infrastructure: Transaction economies in Birmingham and Leicester, UK. *Urban Studies*, 54(6), pp. 1311–1327.

Harrison, R. and Schofield, A.J. (2010). *After modernity: Archaeological approaches to the contemporary past*. Oxford: Oxford University Press.

Hervieu-Léger, D. (2002). Space and religion: New approaches to religious spatiality in modernity. *International Journal of Urban and Regional Research*, 26, pp. 99–105.

Holtorf, C. and Piccini, A., eds. (2009). *Contemporary archaeologies: Excavating now*. Oxford and New York: Peter Lang.

Husband, C. et al., eds. (2014). *Lived diversities: Space, place and identities in the multiethnic city*. Bristol: Policy Press.

Kanmaz, M. (2009). *Islamitische ruimtes in de stad*. Ghent: Academia Press.

Knowles, C. (2013). Nigerian London: Re-mapping space and ethnicity in superdiverse cities. *Ethnic and Racial Studies*, 36(4), pp. 651–669.

Knott, K. (2010). Religion, space, and place. *Religion and Society*, 1(1), pp. 29–43.

Krause, K. (2008). Spiritual spaces in post-industrial places: Transnational churches in North East London. In: M.P. Smith and J. Eade, eds., *Transnational ties: Cities, identities and migrations*. New Brunswick, NJ: Transaction Publishers.

Lanz, S. and Oosterbaan, M. (2016). Entrepreneurial religion in the age of neoliberal urbanism. *International Journal of Urban and Regional Research*, 40, pp. 487–506.

Lefebvre, H. (1968). *Le Droit à la ville*. Paris: Anthropos.

Low, S. (2017). *Spatializing culture: The ethnography of space and place*. London: Routledge.

Lozanovska, M., ed. (2015). *Ethno-architecture and the politics of migration*. London: Routledge.

Lozanovska, M., Levin, I. and Gantala, M.V. (2013). Is the migrant house in Australia an Australian vernacular architecture? *Traditional Dwellings and Settlements Review* XXIV(II), pp. 65–78.

Maskens, M. (2012). Mobility among Pentecostal pastors and migratory 'miracles'. *Canadian Journal of African Studies/La Revue Canadienne des études africaines*, 46(3), pp. 397–409.

Meeus, B., Arnaut, K. and van Heur, B., eds. (2019). *Arrival infrastructures*. London: Palgrave Macmillan.

Merrifield, A. (2012). The politics of the encounter and the urbanization of the world. *City*, 16(3), pp. 269–283.

Mintz, S. (2008). Food and Diaspora. *Food, Culture and Society*, 11(4), pp. 509–523.

Oosterbaan, M. (2014). Public religion and urban space in Europe. *Social and Cultural Geography*, 15(6), pp. 591–602.

Robbins, J. (2004). The globalization of Pentecostal and charismatic Christianity. *Annual Review of Anthropology*, 33(1), pp. 117–143.

Rucka, Z. and Sileghem, H. (2017). *The reversed mission between discourse and practice: Exploring transnational place-making strategies of Afro-Pentecostal churches in the mid-sized city in Belgium (Aalst and Oostende)*. Unpublished MA thesis, University of Ghent. Available at: https://lib.ugent.be/fulltxt/RUG01/002/367/350/RUG01-002367350_2017_0001_AC.pdf

Schillebeeckx, E., Oosterlynck, S. and De Decker, P. (2019). Migration and the resourceful neighbourhood. In: B. Meeus, K. Arnaut, and B. van Heur, eds., *Arrival infrastructures*. London: Palgrave Macmillan.

Sen, A. (2015). Awe and order: Ethno-architecture in everyday life. In: M. Lozanovska, ed., *Ethno-architecture and the politics of migration*. London: Routledge, pp. 151–164.

Sen, A. (2016). Food, place and memory: Bangladeshi fish stores on Devon avenue, Chicago. *Food and Foodways*, 24(1–2), pp. 67–88.

Stad Gent. (2017). Available at: https://gent.buurtmonitor.be

Staeheli, L., Mitchell, D. and Nagel, C. (2009). Making publics: Immigrants, regimes of publicity, and entry to 'the public'. *Environment and Planning D: Society and Space*, 27, pp. 633–648.

Teverson, A. and Upstone, S. (2011). *Postcolonial spaces, the politics of place in contemporary culture*. London: Palgrave.

Valentine, G. (2008). Living with difference: Reflections on geographies of encounter. *Progress in Human Geography*, 32(3), pp. 323–337.

Vyzoviti, S. (2005). The immigrants' place of getting together in downtown Athens. *OASE*, 68, pp. 22–67.

Watson, J. and Cardwell, M. (2005). *The cultural politics of food and eating*. Malden, MA: Blackwell.

Wessendorf, S. (2014). 'Being open, but sometimes closed': Conviviality in a super-diverse London neighbourhood. *European Journal of Cultural Studies*, 17(4), pp. 392–405.

Werbner, P. (2002). The place which is diaspora: Citizenship, religion and gender in the making of chaordic transnationalism. *Journal of Ethnic and Migration Studies*, 28(1), pp. 119–133.

10 Charity, hospitality, tolerance? Religious organisations and the changing vocabularies of migrant assistance in Rome

Luiza Bialasiewicz and William Haynes[1]

In this chapter, we examine how religious organisations engage ideas of 'hospitality' in responding to the needs of irregular migrants in European cities, as well as in contesting the increasingly inhospitable practices of states. Hospitality as a frame for practices of reception and integration of migrant Others has become, in many ways, a byword for tolerance and, just like liberal conceptions of tolerance, a powerful tool in the governance of increasingly multicultural and multireligious European societies (as described by Brown, 2006). We will argue that by drawing attention to some of the ways in which hospitality to migrants is discursively framed and practised in place, we can also better understand the complex ways in which notions of 'tolerance' and 'intolerance' are invoked by a range of actors involved in migrant reception and assistance. While our focus will lie with Rome as a unique site in wider Catholic geopolitics of migration, the points we make here are also, we believe, applicable to numerous other contexts, as the role of faith-based organisations in providing basic assistance to migrant populations continues to grow in Europe and worldwide (Forced Migration Review, 2014).

There is another important reason to examine the work of religious organisations (and the Catholic Church in particular) in practices of institutionalised 'hospitality', and it has to do with the way in which religious *and* secular organisations today increasingly share the very same vocabulary in what is commonly known as 'humanitarian' assistance. Indeed, over the past decade, various scholars examining the changing discourses and practices of humanitarianism have noted how a language of hospitality and charity, often with direct religious connotations, has permeated humanitarian work, both at home and abroad. Didier Fassin's (2012, p. x) work on 'humanitarian reason' is perhaps the most notable in this regard, as "a mode of governing that concerns the victims of poverty, homelessness, unemployment, and exile, as well as of disasters, famines, epidemics and wars – in short, every situation characterized by precariousness". As Fassin argues,

> humanitarianism is the most potent political and geopolitical force of our world. [. . .] In contemporary societies, where inequalities have reached an unprecedented level, humanitarianism elicits the fantasy of a global moral community that may still be viable and the expectation that solidarity may have redeeming powers.

Contemporary humanitarianism is, indeed, profoundly 'moral', framed through a "secular imaginary of communion and redemption" and "a sudden awareness of the fundamentally unequal human condition, and an ethical necessity not to remain passive about it in the name of solidarity – however ephemeral this awareness is, and whatever limited impact this necessity has" (2012, p. xii).

As Fassin but also other scholars examining present-day approaches to migrant assistance[2] have argued, a morally signified language of 'communion and redemption' colours international assistance as well as domestic aid to vulnerable populations: "on both the national and the international levels, the vocabulary of suffering, compassion, assistance, and responsibility to protect forms part of our political life: it serves to qualify the issues involved and to reason about choices made" (2012, p. 2). What is interesting, however, is that alongside these transformations in the language and 'reason' (to pick up Fassin's term) guiding the choices made in granting assistance by state and international actors, we can note a parallel re-framing of the ways in which *also* religious organisations see themselves in the provision of aid, both domestic as well as international. As international bodies such as the UNHCR increasingly speak in a rubric of hospitality, charity, and compassion, religious organisations are taking up international humanitarian actors' language of 'migration management'.

We will return to this perhaps surprising discursive shift in the next section, as we describe the changing institutional and material geographies of reception of recent migrant movements in Europe. Following this overview, we will discuss the role of the Catholic Church specifically in responding to the 2015 'hospitality crisis',[3] noting how the Vatican's (and, in particular, Pope Francis') appeals were translated into responses 'on the ground'. While our focus lies specifically with the Italian context, many of the dynamics we highlight are also identifiable in other EU states: both in the growing role of religious NGOs in providing assistance, but also in emergent tensions between religious and state actors, as the provision of care and refuge to irregular migrants becomes criminalised. Both Italy and France currently possess laws criminalising assistance (and, specifically, 'solidarity' and 'hospitality') to irregular migrants and in many cases, it has been religious organisations and volunteers that have become the focus of state interventions aimed at policing assistance. We conclude the chapter with a brief ethnography of the provision of care to homeless migrants in and around Rome's central train station, Termini, by two religious organisations, the Joel Nafuma Centre and San Vincenzo de Paoli. Examining the work of these two organisations allows us to consider how religious actors use the frame of 'hospitality' to contest and work against those which Maurizio Ambrosini terms the "local policies of exclusion" of Italian municipalities (2013a, 2013b; also Caneva, 2014).

The new geographies of migrant assistance

While the 'summer of migration' of 2015 certainly marked an important shift in the ways in which European states (and especially cities) were forced to respond to the arrival of large numbers of undocumented migrants, fundamental changes in both the institutional as well as material geographies of refugee and migrant reception had been taking place for quite some time already. We can identify five such key shifts that are relevant to our discussion of the European context: (1) a higher proportion of migrants moving to urban and peri-urban settings, rather than refugee camps; (2) movements to destinations further from refugees' origins in conflict zones, particularly – or perhaps most noticeably – by young, unaccompanied migrants; (3) high concentrations of migrants along key transportation corridors and nodes; (4) significant, potentially permanent settlements of migrants near institutional centres in countries of 'first arrival' where they are supposed to register, but not necessarily remain (especially noticeable in Italy and Greece); and (5) the inclusion of non-migratory communities in the provision of assistance, with an increasing 'mixing' of locals and migrants in spaces of care (whether soup kitchens or shelters) (for an overview, see Casas-Cortes et al., 2014).

While previously the reception of migrants and refugees by states as well as international organisations such as the UNHCR was concentrated in camp-like settings, today's spaces of reception are increasingly diffuse. This is also linked to the changing nature of migration flows themselves, which are also more diffuse and characterised by irregular patterns of movement and stoppage. Migrants' journeys today are, more often than not, 'matrioshka' journeys, with multiple stops and periods of permanence interrupting the intended itinerary. It is therefore increasingly difficult to delimit which are countries and sites of 'transit', and which of 'destination'. This necessarily also complicates the provision of assistance, both *en route* and at the points of stoppage that very frequently are in large cities that have become sites of both first reception and longer-term assistance and integration (Agier, 2002; Darling, 2016; de Graauw and Vermeulen, 2016; Delvino, 2017; Sanyal, 2012; Spencer and Delvino, 2019). Moreover, current transformations make it growingly difficult to define and delimit migrants' and refugees' categories of legal status (and thus rights to certain forms of assistance), as individuals both geographically move in and out of formal sites of assistance, but also in and out of types of formal status (see, among others, Allen et al., 2018; Crawley and Skleparis, 2018; Obradovic-Wochnik, 2018).

Humanitarian agencies such as the UNHCR and governments have responded to these changes through enhanced provision of assistance in dispersed sites. But when migrants are located in informal spaces, in temporary accommodation, or find shelter in the quotidian spaces of the city, the nature of responsibility on the part of agencies, governments, residents, and migrants is also transformed. The emergence of a rubric of 'hospitality' rather than of legally specified state responsibility (or of the assurance of basic rights) is also related directly to this shift, as actors who do not, in fact, hold any formal capacity to provide legal status to migrants suddenly find themselves forced to provide at least the most basic forms of assistance, from shelter and food to medical help. The role of city administrations and other urban actors has been extensively discussed by Darling (2010, 2016), Squire and Darling (2013), and Darling and Bauder (2019), examining both formal 'City of Sanctuary' initiatives but also other examples of informal 'urban hospitality' relying on citizen and NGO networks. There is another important set of transformations that must be noted as well, however, as migrant assistance moves into urban environments, for the siting of both reception and first-assistance facilities often creates spatial tensions with the (often also precarious and marginal) local communities within which migrants are placed (Ambrosini, 2013a, 2013b; de Graauw and Vermeulen, 2016; Hoekstra, 2018). As such communities adjust to the presence of migrants, claims are frequently put forward regarding also the right of non-migrant populations to assistance. A variety of European far-right movements have made of such claims one of their rallying cries, not only discursively linking the increasing precarity of domestic urban populations to migrant arrivals but also creating their own practices of urban assistance and hospitality 'for locals only'.[4]

As noted previously, the material shifts in the geographies of assistance and reception of migrants cannot be disentangled from the re-negotiation of responsibilities and obligations to provide such assistance. The deployment of assistance to refugees and migrants has always involved negotiating competing claims to sovereignty, most obviously between states and international organisations (that were called upon to intervene to remedy states' 'failings' in providing for their populations or, more recently, under the rubric of the responsibility to protect, to protect citizens from their own states). Nevertheless, the basic unit on which assistance to migrants and refugees has been predicated since World War II has been the state, just as the recipients of international assistance have also been defined always within a state-bound logic (as abandoned by their states, as necessitating protection from their states). The historical figure of the refugee is, in fact, that of a stateless subject, who has lost the protection of their national-state community and is therefore stripped of rights (Agier, 2011; Hyndman, 2000).

What happens, however, when assistance to refugees and migrants is de-coupled from the state form? When it is no longer just states and inter-state, international organisations that exercise the right to grant rights? Some of these transformations are much longer standing, with the current 'crisis' simply making more visible shifts that have been taking place over decades. One of these, already alluded to earlier, is the profound transformation in the political economy of assistance, mirroring directly the broader re-structuring and roll-back of state functions across both the Global North and South. Such assistance increasingly relies on neo-liberal strategies including public-private partnerships, corporate alliances, and other forms of direct private funding, as well as the direct out-sourcing of functions to private actors, both those on the ground and those operating 'remotely'. Religious and 'faith-based' organisations have played a crucial role in this shift. The Global Compact on Refugees and the Global Compact on Migration presented in 2018 are indicative of this new architecture of international migration governance in which, for the first time, also religious organisations are given a direct role, both in advocacy but also in creating new collaborations with states, private actors, and civil society groups. The International Catholic Migration Commission (ICMC), for example, was one of the groups involved directly in the consultation process leading to the elaboration of the compacts, as too previously in the discussions leading to the New York Declaration for Refugees and Migrants at the United

Nations General Assembly in September 2016 (that mandated the international community of states to develop the compacts) (see ICMC, 2018a, 2018b). What is particularly striking about the role of religious organisations such as the ICMC in the elaboration of the Compacts is not just their newly formalised role, however: it is also how such organisations themselves increasingly discursively frame their support to migrants in the very same language of 'stake-holders', 'opportunities', 'sustainable solutions', 'accountability', and, especially, a focus on the nexus between 'migration and development' in countries of origin (for an example, see the ICMC Strategic Report 2019–2022). As noted in the opening paragraphs, this discursive shift is quite new, although it is unclear to what degree it has also re-shaped the actual practices of Catholic organisations working on behalf of bodies such as the ICMC.

A final important point must be made regarding the effects of the re-structuring of the political economies of migrant and refugee assistance for contests over sovereignty. As has been outlined previously, today negotiations and contesta-tions over the sovereignty of assistance also concern private citizens and non-state organisations bound by different understandings of care and hospitality that do not necessarily correspond to the legal boundaries set by states (and can, indeed, be in direct violation of them). One evident case is precisely that of faith-based organisations such as the many Catholic organisations in Italy (though not only) that have consistently violated national laws that now criminalise assistance to irregular migrants by both providing spaces of sanctuary as well as other forms of assistance and care to migrants whose formal status had as yet not been deter-mined by the state (Ambrosini, 2015).[5] All the while, this is a broader shift that also concerns a variety of NGOs and civil society groups but also local authori-ties, as cities "are becoming increasingly entrepreneurial in developing their own integration philosophies and policies. This has led to cities having markedly dif-ferent approaches to migrant integration, even within the same country" (Scholten and Penninx, 2016, p. 91).

Migrant assistance: humanitarianism, human rights, or 'hospitality'?

We have outlined earlier how the institutional and geographical shifts in the pro-vision of assistance to migrants have also been accompanied by fundamental changes in the understanding and practices of reception and integration, increas-ingly framed not within a language of rights and legal obligations, but within a discourse of 'hospitality'. Indeed, writing already a decade ago, Dikec, Clark and Barnett (2009, p. 2) noted how "reshaped as an ethico-political framework for analyzing the worldly realities of living amongst diverse others, [hospitality] manifests itself at the heart of current debates about immigration, multicultural-ism, and post-national citizenship". What is more, just as the most recent trends in the governance of migration have produced not only increasingly divided regimes of rights to assistance and protection, they have also produced what Ranabir Samaddar has termed a 'calculated hospitality' (Samaddar, 2003), which not only delimits *who* is deserving of hospitality, but also allows for a discharging (and out-sourcing) of *where* the responsibility for providing that hospitality lies (see among others Bulley, 2013; Chauvin and Garces-Mascarenas, 2012, 2014; de Genova and Tazzioli et al., 2016). Again, we wish to draw particular attention to this shift – and to the deployment of the notion of hospitality itself – for the discussions that surround it in today's Europe closely parallel, we believe, the discussions regarding tolerance, with the two frequently entangled (if not confounded): who shall we be hospitable towards? Should there be limits (numerical or other) to hospitality? As Darling (2014, p. 163) notes:

> at a national level, hospitality has been mobilized as a language through which nation-states may present themselves as gracious hosts to new arrivals and as sanctuaries to those seeking refuge. A wide array of countries thus make public statements about their hospitable nature towards refugees and their desire to welcome 'good migrants'. Such a language is always conditioned by the right to select, classify, and limit hospitality.

Hospitality has always been an important philosophical frame in shaping both religious and secular responses to the 'obligations' of migrant reception and assistance. Indeed, many of the admittedly secular international legal frameworks that support the rights of migrants and refugees draw quite directly on both Christian and Islamic theology; many existing relief and migrant assistance organisations, both in Europe and the Middle East originated directly in faith-based organisations. Hospitality towards the stranger is, indeed, a tradition within both Islam and Christianity – as Kaya and Kuyumcuoğlu reminds us in their chapter in this volume. While there are differences in the ways that hospitality is proscribed, it implies the obligation to neighbourliness, even to people who are not immediate neighbours, and to generosity to those who are new or are passing through. It is meant to strengthen society, to build a world in which God's values are paramount, and that brings people closer to God (Siddiqui, 2015; Zaman, 2016). Michael Barnett and Janice Gross Stein (2012) have argued, in fact, that it is important to appreciate the role of religion and religious discourse in shaping notions such as hospitality, even amongst individuals, agencies, and institutions that are putatively secular. Indeed, the taking up of hospitality as a frame for practice in today's (increasingly neo-liberal) migration management is striking, given its theological genealogy "tinged with piety and otherworldly associations", as Dikec, Clark and Barnett (2009, p. 2) note.

The transfer of notions of 'hospitality' to discussions of migrant assistance thus deserves close scrutiny, also in its 'progressive' iterations, considering how closely it is marked by implicitly unequal relations of charity. Highly influential in shaping many migration scholars' conceptualisations of hospitality has been the work of Jacques Derrida (2001; Derrida and Dufourmantelle, 2000). For Derrida, hospitality rests on feelings of altruism that are an *aporia*, an impossibility, in that genuine hospitality to the unknown other embeds its own contradictions. For an individual, it implies that one is master of one's house and is therefore able to act on altruistic impulses and to give what one has to others. Yet such giving, and the demonstration of the power and resources that enable hospitality, also involves the creation of self-identity as compassionate and other-regarding that is, at its heart, selfish. It implies the need for control, order, and the ability to exclude, lest those resources and power are diminished. At a practical level, compassion, altruism, and hospitality are therefore self-limiting and rest on power that is fundamentally at odds with the highest ideals of humanitarianism, as Fassin (2012) argues.

This critique in many ways reflects the critique of the relations of power and abjection underpinning liberal notions of tolerance as articulated by Wendy Brown (2006), and engaged by several chapters in this volume. The relation between contemporary understandings of hospitality and tolerance is closer still, however. Examining the take-up by migration scholars of Derridean notions of hospitality, Barnett (2005, p. 13) reminds us that, for Derrida,

> what he calls 'the laws of hospitality' [are] premised on a logic of unrelinquished mastery over one's own space. This commonsense understanding of hospitality therefore amounts to an ethics of tolerance. And Derrida suggests that tolerance depends on a form of paternalism rooted in mastery and possession, and that it therefore runs counter to the imperative of 'pure' hospitality. This analysis seems to set up a clear opposition between two orders: tolerance, taken as a shorthand for conditional hospitality; and pure, unconditional hospitality.

It is important to therefore fully appreciate the limitations of the rubric of hospitality in migrant reception: as it is conceptualised in its 'commonsense' understandings, hospitality is uni-directional and is not grounded in an unconditional ethics of reciprocity or equality (as too, frequently, liberal understandings of tolerance). Moreover, because hospitality as an ethical practice depends on the willingness – or obligation – of the host to be generous, it is able to sideline an ethical (but also legal) commitment to justice and the rights of refugees and migrants. A reliance on hospitality as a theoretical framework, therefore, can serve to narrow the analytical lens to what states and receiving communities offer to migrants and the politics of such provision, rather than to incorporate migrants fully as agents. It also shifts the discussion away from rights: even in its progressive refractions, a rubric of hospitality is often reduced "to an ethical injunction", as migration scholars such as Squire and Darling (2013, p. 72) have pointed out. Indeed, they (and others) have argued, rather, for "a frame of 'rightful presence' orientated toward exposing the unevenness of relations between guest and host" and, as such, one that does not "aspire to 'inclusion' so much as it does to justice" (2013, p. 72). Only by re-framing discussions of migrant reception and assistance through the rubric of rights and justice rather than 'hospitality', they suggest, allows for the full recognition of migrants as political beings claiming legitimate rights, even though they may be "rights that are 'illegitimate' or 'misplaced' from a statist perspective" (Squire and Darling, 2013, p. 65).

Keeping in mind this critique, we will adopt this concept in our analysis quite selectively in order to (1) highlight the ways in which Catholic leaders and organisations have deployed it strategically as a religious ethics of care for migrant Others and (2) to highlight the tensions between the religious ethic and responsibility to the migrant Other and the sovereignty of the national and local state in policing assistance, in particular through the securitisation of hospitality.

For a 'hospitable Europe'? Contemporary Catholic geopolitics of migration

At the height of the migrant arrivals in the autumn of 2015, in his Sunday 'Angelus' on September 6, Pope Francis appealed to the faithful assembled in St. Peter's Square in the Vatican – and to all European Catholics:

> Faced with the tragedy of tens of thousands of refugees who flee death from war and hunger, and who have begun a journey moved by hope for survival, the Gospel calls us to be 'neighbours' of the smallest and the abandoned, and to give them concrete hope. It's not enough to say, "Take heart. Be patient". [. . .] Therefore, I make an appeal to parishes, religious communities, monasteries and shrines throughout Europe, that they express the Gospel in a concrete way and host a refugee family. A concrete gesture in preparation for the Holy Year of Mercy. May every parish, every religious community, every monastery, every shrine of Europe welcome one family, beginning with my Diocese of Rome.

Pope Francis' appeal was followed, indeed, by action from the Vatican to host a number of refugee families, as well as that of countless other parishes along with individuals and communities across Europe. In a great majority of cases, the hospitality provided accompanied the efforts of international organisations and national and local authorities; in some, however, it directly broke with state and EU policy. The Pope himself travelled to Lesvos at the start of April of 2016, right as the EU-Turkey agreement was going into effect.[6] In violation of the terms of the agreement, he took away with him 12 Syrian refugees, delivered to 'save haven' in the Vatican.[7] As the *Guardian* (2016) commented on the Pope's gesture: "Their arrival brings to about 20 the number of refugees living in the Vatican, which has fewer than 1,000 inhabitants. A similar intake across Europe would see 6 million people given asylum on the continent of 300 million".[8]

During that same visit, in a widely diffused tweet, the Pope chided the EU 'deal-makers' that "refugees are not numbers but people; they are faces, names, stories, and as such they should be treated". European hospitality, he argued, relied, first of all, on this recognition (Catholic News Agency, 2016). While, in the Pope's appeal, this recognition is that of the 'fellow human being in Christ', it is important to note that the question (and tension) of recognition is also at the heart of the notion of 'pure hospitality' as articulated by Derrida. Indeed, we can make a link here to the ways in which Derrida (inspired by the work of Emmanuel Levinas) suggests that "in order to be hospitable, hospitality requires that a guest be greeted, addressed, named as a singular individual", as Barnett (2005, p. 15) argues in his analysis of Derridean notions of hospitality.[9] Citing Naas (2003, p. 159), Barnett recalls, indeed, that:

> hospitality requires that the guest be welcomed as Somebody, not as a serialized nobody. Hospitality requires that one not be indifferent to one's guest. This is not a distinction between an ethical imperative of unconditional welcome and a political imperative to impose conditions, borne out of the empirical necessity to institutionalize rules and regulations. Rather, it is a distinction between two equally compelling ethical imperatives. The imperative to extend unconditional welcome without question and the imperative to impose conditionality on any such welcome by attributing identity are of equal weight within the ethical drama of hospitality.
>
> (2005, pp. 15–16)

Discussions of hospitality inspired by Catholic Social Teaching in many ways engage this very tension (see, among others, Barnett and Stein, 2012), and although we do not have the space here to discuss them further, the question of the 'conditionality' of hospitality has been one repeatedly taken up by the Pope over the past couple of years.

Francis' direct appeals on the question of hospitality to migrants have been a guiding theme of the Papacy, often confronting (if not directly criticising) EU and state leaders. Another important moment in this regard came later that very same spring, in May of 2016, when Pope Francis was awarded the Charlemagne Prize, a prize awarded "for work done in the service of European unification". Flanked by European Commission President Jean-Claude Juncker, President of the European Parliament Martin Schultz, President of the European Council Donald Tusk and EU heads of state, Pope Francis delivered an impassioned address calling upon the assembled leaders to remember Europe's founding purpose as a 'hospitable continent'. Citing his previous Apostolic Exhortation, he appealed for "Europe, rather than protecting spaces" to "be a mother who generates processes" (cf. Apostolic Exhortation *Evangelii Gaudium*, 2016, p. 223). "What has happened to you, the Europe of humanism, the champion of human rights, democracy and freedom?" he asked. "I dream of a Europe" he continued:

> that is young, still capable of being a mother: a mother who has life because she respects life and offers hope for life. I dream of a Europe that cares for children, that offers fraternal help to the poor and those newcomers seeking acceptance because they have lost everything and need shelter. I dream of a Europe that is attentive to and concerned for the infirm and the elderly, lest they be simply set aside as useless. I dream of a Europe where being a migrant is not a crime but a summons to greater commitment on behalf of the dignity of every human being.

A "hospitable Europe", he concluded, "that promotes and protects the rights of everyone, without neglecting its duties to all".

In the months and years that followed the 2015–2016 events, the Vatican as well as well as other leading figures in the Italian Catholic hierarchy have continued to forcefully criticise both EU and Italian state migration policies. Numerous bishops and countless local priests called upon their congregations to directly violate the terms of the various decrees put out by the centre-left Italian government (in office until the spring of 2018). As mentioned previously, this resulted in a number of highly publicised cases of both priests as well as local volunteers for key organisations such as *Caritas* and the *Jesuit Refugee Service* being charged with criminal conduct in 'favouring illegal migration' (Gatti, 2016).

The situation has become further polarised since. The election, in March 2018, of the right-populist coalition government of the *Lega* and the *Movimento Cinque Stelle* brought an immediate hardening of both domestic policies of migrant reception and assistance, and an aggressive crack-down on the patrolling of Italy's maritime and land borders. With the introduction of increasingly migrant-hostile policies, Catholic organisations and activists have found themselves on a direct collision course with the new authorities. The most direct institutional clash came with the introduction, in December 2018, of a new 'Security Decree' (also known as the 'Salvini Decree', after the Minister of the Interior and *Lega* leader, Matteo Salvini). It is a decree with sweeping powers, among other things abolishing with one legislative act the right to 'humanitarian protection' for migrants not eligible for refugee status but who cannot be returned to country of origin. The right to humanitarian protection has been, in the past years, the most utilised instrument to offer the right to remain in Italy, covering over 25% of all instances of awarded protection (as opposed to only 8% of those who were successful in obtaining refugee status, Villa, 2018).

Apart from creating a legal framework allowing migrants to remain on Italian territory, the permits (valid from six months to two years and renewable) previously awarded under the rubric of 'humanitarian protection' allowed migrants to work, and to receive assistance from a network of centres and integration programmes, a widely successful model of 'diffuse reception and integration' called the SPRAR system (in many cases, run directly by or with the assistance of Catholic organisations or cooperatives). With the passing of the Decree, all those benefitting from humanitarian protection would lose those rights – and, moreover, the centres housing them and providing assistance were to be shut down, effective immediately. In the place of humanitarian protection, the Decree specified a new class of special permits can be granted 'at discretion of authorities' to "victims of serious exploitation, victims of domestic violence, people whose country of origin has been hit by disaster, people needing medical treatment and those who have performed acts of high civic value" – the sort of humanitarian 'sorting of rights' decried by Fassin (2012) and others. What is more, the Decree mandated the possibility to deny or revoke asylum for those "who have committed crimes of sexual violence, serious bodily harm, violence against a public official, aggravated robbery, and drug trafficking". Asylum applications could now also be suspended if a person is considered "socially dangerous".

Such a hard-handed approach confirmed Salvini's electoral promise to not only halt migrant arrivals but also to expel all 'undeserving migrants'. Nevertheless, the Decree will not erase illegal migration and illegal migrants from Italian territory: quite the opposite, for the retroactive elimination of the framework and especially of the institutions of humanitarian protection will, literally, produce illegal migrants. It will not only strip thousands of people of the legal right to remain – but also of housing, medical care, and other basic rights. It will literally expel, overnight, thousands of vulnerable individuals and families into the streets. The mayors of several large cities – Milano, Bologna, but also Torino, governed by the *Lega's* coalition partner, *M5S* – reacted immediately, saying that they would not implement the terms of the decree for it risked creating a 'social bomb'. *Caritas* estimated that it will "expose over 140,000 individuals to the risk of extreme poverty, marginality and deviance" and in a circular published the day after the passing of the Decree, termed it a "pathogenic law" (Caritas Roma, 2018). They argued that the decree will not only create illegality – it will also make it much more visible in the physical spaces of Italian cities, with people now thrust into abject conditions forced to survive on the streets.

How can we understand, in this context, the concrete actions of Catholic and other religious organisations to provide assistance and 'hospitality' to irregular migrants, precisely the populations most directly affected by the Salvini Decree? In the following section, we provide a brief overview of some such attempts at a central site, Termini railway station: a site central to migrant arrivals, but also, for its Roman location, central to the Church's attempts to confront the Italian state with its own practices of hospitality in the face of the growing inhospitality of the authorities. The analysis draws upon ethnographic fieldwork carried out by Haynes in November 2017, inspired a long geographical tradition of auto-ethnography as a method for engaging contested public spaces (Low, 2017). An ethnographic approach allows, in particular, for the articulation of more open-ended, performative, and ambivalent understandings of public space (Qian, 2018), and is able to bring together the often divergent and 'muddled' embodied experiences of such sites (Revill, 2013).

Sites of transit, sites of hospitality? European train stations and migrant arrivals

Termini train station is Rome's largest and busiest train station, with almost half a million travellers passing through its corridors every day. Big city train stations have always been key points of both transit and arrival for migrating populations, most recently entering into the European public imaginary as some of the most visible sites of mass arrivals during the 2015–2016 'hospitality crisis'. Much of this has to do with stations' logistical role as crucial connecting points or nodes on migratory routes (generally from entry points on the Mediterranean and along the 'Balkan route', moving north-westwards), though stations have also had a powerful symbolic role in 'emplacing' representations of the 'crisis'. Countless images of arrivals, departures, encampments, and evictions at European train stations proliferated in the mass media from the summer of 2015 onwards, not only highlighting the material importance of these spaces in the trajectories of migrating people, but also the co-presence and intertwinement of the lives of migrants and the everyday lives and movements of Europeans, with scenes of desperation and daily commuter traffic taking place side by side. Train stations became sites of acts of solidarity by individuals and organisations, but also acts of aggression, whether at Vienna's Hauptbanhof and Westbanhof (Viderman and Knierbein, 2018), Budapest's Keleti Station, or other points further north and west (Collyer and King, 2016). International news reports depicted highly emotional scenes of halted departures and desperate encampments, whether on the railway tracks at the Greek-Macedonian border at Idomeni (Smith, 2016), or those of migrants stranded at Belgrade's central station, making fires from garbage to keep warm (Dickerman, 2017). As a second-order site in the migration corridors of the Mediterranean and Western Balkans, Termini did not receive the sort of international media attention that focussed on the places of first arrival or transit along the 'Balkan Route'. Nevertheless, its population of homeless migrants, some merely in transit, some much more permanent, has continued to grow already since 2011 (see Accorinti and Wislocki, 2016). In their report of 2018, Médecins Sans Frontières (2018) estimated that Termini was occupied by close to 100 homeless migrants each night, with the actual figures most likely double that estimate. Termini, like many other European central railroad stations, has become part of what Rygiel (2011) has identified as an emerging global network of spaces of exception and encampment. Yet while scholars like Agier (2011) have highlighted the ways in which camps have become increasingly permanent and urbanised, noting that the 'city is in the camp', Darling (2016) inverts this phrase, noting it is rather the camp that now haunts European cities, as 'humanitarian spaces' appear within and alongside the 'normal' spaces of urban life. Observing the multitude of figures sleeping on makeshift cardboard beds at the front entrance to Termini station, right in the midst of early morning commuter traffic affirms this statement.

Train stations have always been what Löfgren (2008) terms a *terraine vague*, open public spaces that allow those on the margins the possibility of finding the necessary resources for survival, and that provide anonymity and limited social control: spaces where one can literally 'disappear into the crowd' (Castrignanò, 2004). Stations provide a modicum of shelter, possibly access to bathroom facilities and the chance of obtaining discarded food or the handouts of travellers. However, the redevelopment of city centres and an increasing privatisation of public spaces (Smith and Low, 2006) has transformed the role of stations as well. Like many other central train stations in Europe, Termini has undergone significant redevelopment works over the past two decades, transforming the station into a kind of urban square, "rich in services and shopping opportunities" (Carminucci, 2011, p. 68). The commercialisation of the station has also been accompanied, unsurprisingly, by increasing securitisation, with increasing architectural barriers as well as physical patrolling of the Termini space to regulate the presence of those who are 'out of place' in the station-as-shopping mall.

Religious place-making at Termini

Nevertheless, before their contemporary transformation into privatised spaces of consumption, large train stations (like other major public structures) have often strongly resembled religious buildings. Indeed, according to train historians Richards and Mackenzie (1986, p. 11), the symbolic role of stations was "essentially ecclesiastical". From the origins of modern train stations in Europe during the 19th century until the present day, central stations were frequently referred to in almost spiritual language: 'cathedrals of steam' or 'temples of transportation'. The parallels were not just semantic: the now-demolished Great Hall at Euston Station in London was modelled on Saint Peter's Basilica in Rome, Antwerp Centraal features a cathedral-like soaring stone atrium and is crowned by a huge dome, whilst Bilbao-Abando boasts a large and ornate stained-glass window in its upper hall celebrating the industrial achievements of the time. The 'religious' dimension of stations was made present not only in terms of architecture and material design, however, but also in the functions assigned to their spaces. As Löfgren (2015, p. 85) points out, "the station may seem like a very stable bricks-and-mortar monument, but it is really built by all the comings and goings, as well as the very diverse tasks, motives and mental luggage dragged into it". Central train stations, like directly religious spaces, were always relatively open spaces, indiscriminate in who they admit. Much like church spaces, they also frequently provided the basic requisites of shelter (though of course not in the more formal sense of the sanctuary provided by many churches – see Neufert, 2014).

In the case of Termini specifically, the religious links are much more direct. Like much of Rome's urban fabric, it is inseparable from the history of the Papacy and the Catholic Church. It was, in fact, Pope Pius IX, the last ruler of the Papal States, who initiated plans to bring the railroad to Rome in order to 'modernise' the city in the 1850s, as Weststeijn and Whitling (2017) note in their comprehensive history of the station. After Rome was captured by Italian forces in 1870 and was subsequently made into the new Italian state's capital, the station became the focus of the nation-building strategies (Painter, 2005). Termini was to be a symbol both of Rome's status as the meeting-place of antiquity and modernity, as well as the central locus of the new Italian state. The station was eventually completed in 1873 to much fanfare (Weststeijn and Whitling, 2017). The Fascist state similarly made Termini a focal point of its urban and national imaginaries. Termini itself was restructured, but especially made a key node in the new network of subway lines that was meant to connect the centre of Rome to the new district being built around the Esposizione Universale di Roma 1942, intended as a model Fascist city (today's EUR district) (d'Eramo, 2017).

Following the war, the station was rebuilt in 1950 and stands today as one of Rome's most iconic modernist structures (see Insolera, 1993). A passage from our 2017 ethnography reveals first impressions of the station's arrival hall:

> After rising from the claustrophobic depths of the station, the scale of the main hall is monumental. Fronted by high glass walls, the room is vast,

topped by an impressive vaulted ceiling. Whilst made of raw concrete, this ceiling hints at a direct link to Ancient Rome. It is essentially an angular, 20th Century interpretation of a barrel vault, with the uniformity of its concrete girders resembling colonnades. Beneath the ceiling, this large open space offers a mix of familiar sights and sounds. It is filled with the bright lights of advertisements: rotating billboards showcasing new the new menu at McDonalds and 'exclusive' deals from affordable mobile phone networks. Slicing through the hum of conversation, is the bleeping of ticket machines. It smells like fast food and coffee, with passing bursts of perfume and sweat.

In the late 1990s, Termini became once again the focus of redevelopment plans, this time linked not to the remaking of the Italian state but to the Great Jubilee of 2000, the largest celebration in the history of the Catholic Church to date. While the Jubilee was used as an opportunity to refashion a number of Rome's most prominent landmarks (and not just those directly linked to the Church), Termini was afforded a special place: both for its logistical importance in facilitating the arrival of the millions of pilgrims expected, but also as a symbolic site of entrance into the Holy City. In his account of the redevelopment of Rome in the late 1990s and early 2000s, D'Eramo (2017) describes the unstated presence of the Vatican in what he terms a 'clerico-company town':

> If you ask politicians in Rome for some glimpse into the relationship between the city and the Vatican, the answer resembles the way Americans talk about race relations and Indians speak of caste: Vatican, race and caste are problems of the past, which at one point were very serious, but which now are 'resolved, or almost'. But from a planetary standpoint, the relevance of Rome consists all but exclusively of the Vatican. The Vatican is the company and Rome is the town.

The role of the Church is therefore not only symbolic – it is very material as well, particularly in the realm of real estate:

> A quarter of the entire real estate in Rome belongs to the Church. We know for sure that Propaganda Fide owns 725 buildings in the city, while APSA (*Amministrazione del patrimonio della sede apostolica*) possesses 5,050 apartments. Other sources report 50 monasteries, more than 500 churches, 22 convents and 400 buildings that include houses, seminaries, oratories and about 40 colleges. And this patrimony keeps growing; in the city of Rome alone, the Church receives approximately 8,000 bequests each year. In Italy, almost all approved private health care, in large hospitals or clinics, is controlled by the Vatican, and virtually all private education is Catholic. Until about 1980, the Vatican was a direct economic agent on the market, active in real-estate speculation, in keeping with a tradition that began with the unification of Italy, when financiers connected to the Church began to sell off some of the vast holdings of religious orders in the city and its surroundings to profiteers.

> (d'Eramo, 2017, p. 17)

We note these figures to highlight the temporal (if not directly economic-speculative) role of the Church in the Roman context – important also in considering its role in the provision of assistance to migrants, including housing.

A last important passage pertinent specifically to the Church's relation to Termini came in 2006, when the station was renamed, following the celebrations of the Great Jubilee, *Termini-Giovanni Paolo* ('John Paul II') (Brioni, 2017, p. 451). In 2011, the connection was further emphasised with the installation of a bronze statue of Pope John Paul II in Piazza dei Cinquecento (the large transport hub right in front of the station), coinciding with the Pope's beatification by his successor Benedict XVI (BBC, 2011). Sculpted by Oliviero Rainaldi and entitled *'Conversazioni'* ('Conversations'), the statue sparked controversy, mostly for the figure's apparent unlikeness to the previous Pope. The monument was eventually redesigned, but for all its aesthetic flaws, it underlines the symbolic influence of the Church on the station and surrounding landscape. The imposing nature of the statue is detailed in our ethnography:

> at the far-end of the square, at its furthest point from the station, I encounter something unexpected. Here, looming 5 or 6 metres overhead is a gigantic statue of a cloaked man. Lit by streetlamps, it is made from oxidised copper, identifiable by its verdi-gris colour, the same as on exposed church roofs or spires. The man has a familiar, yet swollen, misshapen head and a wry smile. The peculiar head sits upon a cloaked body, entirely hollow within. With one

Figure 10.1 Statue of Pope John Paul II, Piazza dei 500, Rome

arm outstretched, the body is opened in a gesture of welcome, as if to greet those wandering by in the spirit of compassion or welcome.

The Church is present in countless other ways in the immediate surroundings of Termini. The Church of Santa Maria degli Angeli dei Martiri, lies north-west of Piazza dei Cinquecento, and besides being considered Michelangelo's swansong, it holds special importance for it was built over much of the Baths of Diocletian, which provide the station with its name from *terme*, or 'baths'.

A space of (in)hospitality?

Alongside its material and symbolic connection with the Church, Termini (like other central train stations) has also played a key symbolic role in Rome as an imagined space of danger, transgression, and sin, rather than piety. Brioni (2017, p. 448), describing five decades of cinematic representations of Termini and its surrounding neighbourhoods, notes how the station has long been presented as a dangerous space, a space of sexual promiscuity, as well as a site where various criminal activities take place, from the peddling of stolen goods, to drug dealing, and male prostitution. But the spectre of the 'dark side' of Termini extends beyond the station itself and into the surrounding neighbourhood of Esquilino, where adult movie houses were long the sites of transgressive sexual exchanges, as too the ruins of the ancient baths (Restivo, 2002). In recent years, numerous cases of murder, rape, and sexual violence occurring at Termini or its immediate vicinity captured the attention of the Italian and international press, increasingly linked to the growing migrant presence in the area.

The station has become, in fact, the focus of various initiatives to secure both its premises and the surrounding landscape, including interventions to discourage antisocial behaviour and unwanted visitors. The use of defensive architecture and defensive urban furniture (benches, railings etc.) has been long documented as a tool to control public space and more specifically to restrict the access of unwanted groups such the homeless and migrants (Bergamaschi et al., 2014; Davis, 1990; Mitchell, 1998). In similar ways, music has also been instrumentalised as a strategy to discourage the presence of certain groups. As Frers (2006) and Di Croce (2017), among others, have noted, the use of music and loud sounds in train stations is a common tactic to discourage permanence, challenging the notion of stations as public spaces able to offer (at least moments of) hospitality. Through the strategic use of noise and other interventions, stations can be re-shaped from spaces of publicness into spaces of discipline and control, 'moving on' if not directly expelling those seeking refuge. Löfgren (2008, p. 345) describes, for instance, the use of loud music at Copenhagen Central Station, summed up by a local newspaper as: "Verdi and Wagner [now] keep the junkies away". In the pedestrian tunnels leading to Termini's Metro platforms, forms of spatial exclusion are evident also on a sonic level. Walking through the station's underground tunnels, the ethnography notes:

> down here, there are absolutely no chairs to rest on, or benches that might be used for sleeping. The only surface is the cold, hard marble floor that reflects

the harsh light of the wall lamps. There is also loud choral music blasting from the speakers, playing to a non-existent audience and echoing through the empty passage.

In what is increasingly an un-hospitable space – both for its commercialisation and privatisation described previously, but also the various tactics aimed at exclusion noted earlier – we can nevertheless observe a variety of acts of hospitality in and around the station. The role of religious groups is key in this regard, and the next paragraphs will describe some of their initiatives. Although the two best-known and most active Catholic organisations that provide assistance to migrants in Rome (and in Italy more generally) are *Caritas* and the *Centro Astalli* (Accorinti and Wislocki, 2016), our focus will lie with the activities of two less-known groups, the *Joel Nafuma Refugee Center* and *San Vincenzo de Paoli*, both particularly active in the areas around Termini.

Società San Vincenzo de Paoli is a Catholic charitable association that operates worldwide, named after Saint Vincent de Paul, a French Catholic priest born in the 16th century who dedicated himself to serving the poor and was venerated for his compassion, humility, and generosity. Globally, the Society of Saint Vincent de Paul had 800,000 members in 2012, according to its website. It possesses branches in multiple Italian cities, though its national office is in Rome. During the ethnographic work at Termini, a group of volunteers from the *Società* was handing out food and drinks late in the evening:

There are about 30 figures standing in an opening between the stone pines that line the square and a row of large potted plants. Half of them are wearing yellow high-visibility jackets. Their backs read 'San Vincenzo de Paoli' under a symbol of the ichthys. The others are the homeless of Termini, now gathered here. I can see those who rose from the depths of the station, and indeed those to whom the blanketed spaces outside the station belong. Some sit together on large potted plants and benches, whilst others take their food not far away and eat alone, under a tree or sitting on the stone handrail at the metro entrance. The food appears to be rice or pasta, and is wolfed down hungrily. Beside the scene, a police officer stands guard, and two Carabinieri cars are parked, with their occupants observing the square.

San Vincenzo de Paoli runs canteens, shelters, and dormitories throughout Italy. As Termini has itself become a temporary space of shelter, their activities have been extended there as well. Indeed, their Rome mission provides hot food and drinks to the homeless inhabitants of the station, both migrant and 'local', on a nightly basis. Its volunteers are recognisable by the organisation's logo emblazoned on their neon jackets – the ichthys fish – the ancient symbol of Christianity and evoking Jesus' feeding of the multitude. The Roman branch has, indeed, several videos on YouTube, demonstrating its various activities: preparing large quantities of food and serving meals in church halls to groups of the poor, as well as providing sports and other activities for children, and long-term 'integration assistance' to migrants in Valmontone, right outside of the city limits.

The other organisation with a notable presence at the station is the *Joel Nafuma Refugee Centre*, a day-centre that operates in the crypt of St Paul's Within the Walls Episcopal Church, less than one kilometre from the Piazza dei Cinquecento. According to its website, it aims at providing a support network for refugees and asylum seekers as well as "offering encouragement and the means to build new lives". It is open daily from 9 AM to 2 PM, providing a daytime space where homeless migrants can spend time each day. It is the only day-centre for migrants in Rome, although inevitably, its beneficiaries must find their own way once it closes for the afternoon. The JNRC offers several different services for its 'guests' including breakfast; toiletries and clothing donations; language classes in Italian, English, and German; medical help, counselling, and art therapy; and a 'peace garden'. Despite an increase in exposure since the height of migrant arrivals in 2015, it has assisted the arrivals of migrants since the 1980s, when the then-rector of the Church invited Reverend Joel Nafuma, a Ugandan priest and refugee, to take up a formal ministry in Rome. It is not connected to the government, although it is partnered with a variety of organisations around the world such as UNHCR and various Protestant church groups. Most of its workers are volunteers. While they encourage 'living through Christian values', they welcome guests from all religious backgrounds. The JNRC has an active social media presence, and its Instagram regularly features images not only highlighting the activities taking place at the centre, but also raising awareness of the desperate situation at Termini.

Although undoubtedly important, the role played by these two organisations is limited in both space as well as time: while a homeless migrant at Termini station may be able to obtain a cooked meal from *San Vincenzo de Paoli*, or spend the day in the spaces of the *Joel Nafuma Refugee Centre*, once these moments and spaces of assistance are no longer available, they will be unable to find shelter in the station's tunnels as the piped-in choral music is too loud and security forces push unwanted visitors way. The moments of 'hospitality' (for it is in such language that the assistance of both organisations is presented) give way to a securitised, inhospitable space. The temporary presence of religious actors at the station could, indeed, be seen represents the *aporia* of hospitality in Rome. This is true in a number of ways: first, because while the provision of hospitality by the two groups may provide relief for migrants' most urgent and desperate needs, it does not provide a long-term solution. Second, while not wanting to cast doubt on the charitable intentions of the groups and the volunteers working with them, their highly visible performances of hospitality in some way serve to assuage the consciences both of passers-by ('at least someone is doing something for these people') but also in part of the local authorities. Indeed, the work of religious organisations such as these contributes in unfortunate ways to the discharging of responsibility of the local and national state (again, with the trope of 'at least someone is doing something'). What is more, the assistance provided by such groups is usually tolerated by even those authorities whose decrees it explicitly violates, for it is the work of such organisations that allows the problems of a growing homeless population from getting out of hand – as Accorinti and Wislocki noted in their report of 2017, their assistance is seen as key to preventing threats to both public health and public security.

Nevertheless, with the signing of the new Salvini Decree in December 2018 severely limiting the provision of any sort of assistance to irregular migrants, this situation is very likely to change. Indeed, the new *Lega* leadership has become increasingly critical of not just religious organisations engaged in assistance to migrants, but of the Vatican itself, and especially Pope Francis. Many *Lega* politicians (including the current Minister of Families, Lorenzo Fontana) have very close ties to ultra-conservative Catholic currents, and have long been vocal in their criticisms both of the new Pope, but also of leading Catholic 'social actors' like *Caritas* precisely for their 'mistaken hospitality' that, in the words of Fontana, risks 'submerging Europe'.[10]

This tension is not new: writing in 2014, Caneva noted how the *Lega* at the local level has taken up the mantle of the defender of Catholic values, presenting itself as the 'true' exponent of Italian Catholicism as both a 'civil religion' and marker of Italian national identity – in direct "opposition to Catholic ecclesiastic authorities (e.g. Caritas, a Catholic pastoral body that promotes charity) that preach openness and tolerance towards Muslims and other religious groups" (Caneva, 2014, p. 384). With the Lega's entry into government – and indeed its dominance of the governing coalition on matters related to migration – this discourse has become even more pronounced. With the Salvini Decree, the Lega has unleashed a veritable 'war on hospitality', as various commentators have termed it. Writing in the weekly magazine *L'Espresso*, veteran journalist Fabrizio Gatti (2019) termed it 'Caccia a chi aiuta' ('a hunt against those who assist'), describing the political utility of the Decree:

> migrants should, in fact, go to sleep on the streets, they should reduce stations like Termini into dormitories for the desperate. In this way, those who support this government can see them, and can be ever more afraid. And if one of these desperate masses, pushed by hunger or by anger, breaks the law – even better.

By stripping them of even the most basic forms of assistance, irregular migrants on Italian territory should be reduced to such an abject condition as to be undeserving of hospitality.

In closing, we wish to reflect on an episode that took place precisely at Termini, right as this volume was going to print. On April 23, 2019, following a fight between two homeless men at the station, one stabbed the other in the throat: the victim, a citizen of Georgia, declared initially to the authorities that it was "because I was wearing a crucifix". Although the man subsequently changed his account of the circumstances of the attack, his accusation was immediately catapulted to the front pages of the local and national papers: "homeless man stabbed because he was wearing a crucifix" declared Italy's second highest circulation daily newspaper, the centre-left *La Repubblica* (Savelli, 2019). The identification of the attacker as a Moroccan with a prior criminal record further fed the media frenzy, as the arresting authorities charged the man with "attempted homicide, aggravated by religious hatred". What was even more grave in this episode, however, was the way it was immediately picked up by the Minister of the Interior (and *Lega* leader) Matteo Salvini. Just hours after the attack, Salvini called for a special meeting of the Council of Ministers on "security, terrorism, Islamic extremism and immigration", and declared that he would "instruct all prefects and police commissioners to at once increase controls in all areas where Muslims gather, in order to prevent acts of violence against innocent citizens" (cited in Savelli, 2019; see also Frignani, 2019 on Italy's highest selling daily, *Corriere della Sera,*). The immediate re-framing as 'religiously motivated' of what was, most likely, simply a desperate act between two equally desperate individuals, abandoned to fend for themselves in the desperate spaces of Termini, forcefully reminds us of the power of discourses of religious intolerance in today's Europe. Using the rubric of 'religious hatred',[11] an immediate link is created in the popular and political imaginary not just between migration and marginality but also between migration and (highly racialised) categories of religious 'peril'. Through the coupling of 'security, terrorism, Islamic extremism and immigration', Italy's and Italians' safety becomes reliant on the proper policing both of migration – and of religion.

Notes

1 The initial ideas for this chapter were developed during a research stay in Rome in November 2017, as part of a graduate course on 'Cities, Borders and Identities' based at the Royal Netherlands Institute in Rome (KNIR). The authors would like to thank the KNIR and its director Harald Hendrix for hospitality and support. Luiza Bialasiewicz would also like to thank Lynn Staeheli for ongoing conversations regarding the changing spatialities of humanitarianism and the role of religious organisations: some of the considerations developed here grew out of a joint (alas unsuccessful) grant application to the National Science Foundation on 'Engaging Humanitarianism' in 2016.

2 See also, among others Agier (2011), Feldman and Ticktin (2013), and Ticktin (2011), as well as Hyndman (2000).

3 As migration scholars and activists have suggested the events of 2015–2016 be more appropriately called.

4 The Rome-based new-Fascist movement *CasaPound* has made the provision of 'hospitality' for Italians facing eviction one of their best-known strategies: see Bialasiewicz and Stallone (2019).

5 For a discussion of one highly publicised case involving three Caritas volunteers, see Gatti (2016).

6 The agreement specified that after the date of March 20, 2016, all migrants attempting to reach Greece from Turkey were to be halted and returned; for a fuller discussion of the agreement and its appeals to a humanitarian logic, see Bialasiewicz and Maessen (2018).

7 Although as the Pope declared to reporters on his flight back from Lesvos: "there is no political speculation [here]. Bringing these refugees away is a humanitarian thing. It was an inspiration from a week ago that I immediately accepted, because I saw that it was the Holy Spirit who was speaking" (Catholic News Agency, 2016).

8 A more critical analysis of the Pope's gesture is provided by Pallister-Wilkins (2016): https://4mpodcast.com/2016/05/09/the-papacy-paternalistic-as-ever-a-response-to-anja-karlsson-franck/.

9 See also the discussion in Popke (2006).
10 Fontana published a book in 2018 entitled *The Empty Cradle of Civilization* (with a preface by Matteo Salvini) warning against Europe's demographic collapse, faced with a combination of massive migrant arrivals and a decline of traditional family structures.
11 As the centre-right Rome daily *Il Messaggero* (2019) announced in its headline, paired with the assertion: "Salvini raises security alert".

References

Accorinti, M. and Wislocki, A-S. (2016). Housing for Asylum Seekers and refugees in Rome: Non-profit and public sector cooperation in the reception system. IRPPS Working Paper, 86/2016.

Agier, M. (2002). Between war and city: Towards an urban anthropology of refugee camps. *Ethnography*, 3, pp. 317–341.

Agier, M. (2011). *Managing the undesirables: Refugee camps and humanitarian government*. Cambridge: Polity Press.

Allen, W. et al. (2018). Who counts in crises? The new geopolitics of international migration and refugee governance. *Geopolitics*, 23(1), pp. 217–243.

Ambrosini, M. (2013a). 'We are against a multi-ethnic society': Policies of exclusion at the urban level in Italy. *Ethnic and Racial Studies*, 36(1), pp. 136–155.

Ambrosini, M. (2013b). *Irregular immigration and invisible welfare*. Basingstoke: Palgrave.

Ambrosini, M. (2015). NGOs and health services for irregular immigrants in Italy: When the protection of human rights challenges the laws. *Journal of Immigrant and Refugee Studies*, 13(2), pp. 116–134.

Barnett, C. (2005). Ways of relating: Hospitality and the acknowledgement of otherness. *Progress in Human Geography*, 29, pp. 5–21.

Barnett, M. and Gross Stein, J., eds. (2012). *Sacred aid: Faith and humanitarianism*. Oxford: Oxford University Press.

BBC (2011). *New statue of Pope John Paul II sparks anger in Rome*. [online] *BBC News*. Available at: https://www.bbc.com/news/world-europe-13475350 [Accessed 10 Dec. 2018].

Bergamaschi, M. et al. (2014). The homeless and public space: Urban policy and exclusion in Bologna. *Revue Interventions Économiques,* 51, pp. 1–20.

Bialasiewicz, L. and Maessen, E. (2018). Scaling rights: The Turkey 'deal' and the divided geographies of European responsibility. *Patterns of Prejudice*, 52(2/3), pp. 210–230.

Bialasiewicz, L. and Stallone, S. (2019). Focalizing new-fascisms: Right politics and integralisms in contemporary Italy. *Environment and Planning C: Politics and Space*, https://doi.org/10.1177/2399654419871303.

Brioni, S. (2017). A station in motion: Termini as heterotopia. *Italian Studies,* 72(4), pp. 443–454.

Brown, W. (2006). *Regulating aversion: Tolerance in the age of identity and empire*. Princeton, NJ: Princeton University Press.

Bulley, D. (2013). Conducting strangers: Hospitality and governmentality in the global city. In: G. Baker, ed., *Hospitality and world politics*. Basingstoke: Palgrave Macmillan.

Caneva, E. (2014). Intolerant policies and discourses in Northern Italian cities. *Journal of Immigrant and Refugee Studies*, 12(4), pp. 383–400.

Caritas Roma (2018). Decreto sicurezza: Una legge patogena, inutile e dannosa. *Caritas Roma* 29 Nov. Available at: http://www.caritasroma.it/2018/11/decreto-sicurezza-una-legge-patogena-inutile-e-dannosa/

Carminucci, C. (2011). Models of social action and homeless support services mapping for some major European train stations. *European Journal of Homelessness*, 5(2), pp. 63–80.

Casas-Cortes, M. et al. (2014). New keywords: Migration and borders. *Cultural Studies*, 29(1), pp. 55–87.

Castrignanò, M. (2004). *La città degli individui: Tra crisi ed evoluzione del legame sociale.* Milano: FrancoAngeli.

Catholic News Agency (2016). Full text of Pope Francis' in-flight interview from Lesbos to Rome. 16 Apr.

Chauvin, S. and Garces-Mascarenas, B. (2012). Beyond informal citizenship: The new moral economy of migrant illegality. *International Political Sociology*, 6(3), pp. 241–259.

Chauvin, S. and Garces-Mascarenas, B. (2014). Becoming less illegal: Deservingness frames and undocumented migrant incorporation. *Sociology Compass*, 8(4), pp. 422–432.

Collyer, M. and King, R. (2016). Narrating Europe's migration and refugee 'crisis'. *Human Geography: A New Radical Journal*, 9(2), pp. 1–12.

Crawley, H. and Skleparis, D. (2018). Refugees, migrants, neither, both: Categorical fetishism and the politics of bounding Europe's migration 'crisis'. *Journal of Ethnic and Migration Studies*, 44(1), pp. 48–64.

Darling, J. (2010). A city of sanctuary: The relational re-imagining of Sheffield's asylum politics. *Transactions of the Institute of British Geographers*, 35, pp. 125–140.

Darling, J. (2014). From hospitality to presence. *Peace Review*, 26(2), pp. 162–169.

Darling, J. (2016). Forced migration and the city: Irregularity, informality and the politics of presence. *Progress in Human Geography*, 41(2), pp. 178–198.

Darling, J. and Bauder, H. (2019). *Sanctuary cities and urban struggles: Rescaling migration, citizenship, and rights.* Manchester: University of Manchester Press.

Davis, M. (1990). *City of quartz: Excavating the future in Los Angeles.* New York: Verso.

d'Eramo, M. (2017). The not so eternal city. *New Left Review*, July–Aug., 77–103.

de Genova, N. and Tazzioli, M. et al. (2016). Europe/crisis: New keywords of 'the crisis' in and of 'Europe'. *Near Futures Online 1 'Europe at a Crossroads'*. Available at: http://nearfuturesonline.org/europecrisis-new-keywords-of-crisis-in-and-of-europe/

de Graauw, E. and Vermeulen, F. (2016). Cities and the politics of migrant integration: A comparison of Berlin, Amsterdam, New York City and San Francisco. *Journal of Ethnic and Minority Studies*, 42(6), pp. 989–1012.

Delvino, N. (2017). *European cities and migrants with irregular status: Municipal initiatives for the inclusion of irregular migrants in the provision of services.* Oxford: COMPAS.

Derrida, J. (2001). *On cosmopolitanism and forgiveness.* London: Routledge.

Derrida, J. and Dufourmantelle, A. (2000). *Of hospitality* (R. Bowlby, trans.). Stanford, CA: Stanford University Press.

Di Croce, N. (2017). Sonic territorialisation in motion: Reporting from the homeless occupation of public space in Grenoble. *Ambiances*, 3, pp. 1–14.

Dickerman, K. (2017). *Stuck at Belgrade station: Photos show deplorable conditions migrants and refugees are left living in.* [online] *The Washington Post.* Available at: https://www.washingtonpost.com/news/in-sight/wp/2017/05/15/stuck-at-belgrade-station-photos-show-deplorable-conditions-migrants-and-refugees-are-left-living-in/ [Accessed 17 May 2018].

Dikeç, M., Clark, N. and Barnett, C. (2009). Extending hospitality: Giving space, taking time. *Paragraph*, 32(1), pp. 1–14.

Fassin, D. (2012). *Humanitarian reason: A moral history of the present*. Berkeley, CA: University of California Press.

Feldman, I. and Ticktin, M., eds. (2013). *In the name of humanity: The government of threat and care*. Durham, NC: Duke University Press.

Forced Migration Review (2014). *Special issue on 'faith and responses to displacement'*. Issue 48, Nov. Oxford: University of Oxford Refugee Studies Centre.

Frers, L. (2006). Pacification by design: An ethnography or normalisation techniques. In: H. Berking et al., eds., *Negotiating urban conflicts: Interaction, space and control'*. Bielefeld: Transcript, pp. 249–262.

Frignani, R. (2019). Roma, accoltellato a Termini 'per il crocefisso': giallo sulla lite tra due clochard. *Corriere della Sera*, 23 Apr.

Gatti, F. (2016). Accompagnano i profughi alla Caritas: a Udine tre volontari rischiano il processo. *L'Espresso*, 13 June.

Gatti, F. (2019). Caccia grossa a chi aiuta. *L'Espresso*, 29 Jan.

Guardian (2016). Pope Francis hailed as saviour by Syrian refugees taken in by Vatican. 17 Apr.

Hoekstra, M.S. (2018). Governing difference in the city: Urban imaginaries and the policy practice of migrant incorporation. *Territory, Politics, Governance*, 6(3), pp. 362–380.

Hyndman, J. (2000). *Managing displacement: Refugees and the politics of humanitarianism*. Minneapolis: University of Minnesota Press.

ICMC (2018a). The global compacts and the role of *the* church. *International Catholic Migration Commission, Press Release*, 7 Mar.

ICMC (2018b). The global compact for migration: What happened? And what's next? *International Catholic Migration Commission, Press Release*, 19 Dec.

Insolera, I. (1993). *Roma moderna: Da Napoleone I al XXI secolo*. Rome: Laterza.

Joel Nafuma Refugee Center (2018). *About us*. [online] Available at: https://jnrc.it [Accessed 29 Feb. 2019].

Löfgren, O. (2008). Motion and emotion: Learning to be a railway traveller. *Mobilities*, 3(3), pp. 331–351.

Löfgren, O. (2015). Sharing an atmosphre: Spaces in an urban commons. In: *Urban Commons: Rethinking the city*. Oxford: Routledge, pp. 68–91.

Low, S. (2017). *Spatializing culture: The ethnography of space and place*. London: Routledge.

Médecins Sans Frontières/Medici Senza Frontiere (2018). *Fuoricampo: Insediamenti informali*. Rome: Medici Senza Frontiere.

Mitchell, D. (1998). Anti-homeless laws and public space: Begging and the First Amendment. *Urban Geography*, 19(1), pp. 6–11.

Messaggero, Il (2019). Termini, lite tra clochard per un crocifisso. PM: 'odio religioso'. Salvini alza la sicurezza. *Il Messaggero*, April 23.

Naas, M. (2003). *Taking on tradition: Jacques Derrida and the legacies of deconstruction*. Stanford, CA: Stanford University Press.

Neufert, B. (2014). Church Asylum. *Forced Migration Review*, 48, pp. 36–38.

Obradovic-Wochnik, J. (2018). Urban geographies of refugee journeys: Biopolitics, neoliberalism and contestation over public space in Belgrade. *Political Geography*, 67, pp. 65–75.

Painter, B.W.J. (2005). *Mussolini's Rome: Rebuilding the Eternal City*. Basingstoke: Palgrave.

Pallister-Wilkins, P. (2016). The Papacy, paternalistic as ever: A response to Anja Karls-son Franck. *4M Podcast*. Available at: https://4mpodcast.com/2016/05/09/the-papacy-paternalistic-as-ever-a-response-to-anja-karlsson-franck/.

Popke, J. (2006). Geography and ethics: Everyday mediations through care and consumption. *Progress in Human Geography*, 30, pp. 504–512.

Qian, J. (2018). Geographies of public space: Variegated publicness, variegated epistemologies. *Progress in Human Geography*. https://journals.sagepub.com/doi/10.1177/0309132518817824

Restivo, A. (2002). *The cinema of economic miracles: Visuality and modernization in the Italian art film*. Durham, NC: Duke University Press.

Revill, G. (2013). Points of departure: Listening to rhythm in the sonoric spaces of the railway station. *The Sociological Review*, 61(1), pp. 51–68.

Richards, J. and Mackenzie, J. (1986). *The railway station: A social history*. Oxford: Oxford University Press.

Rygiel, K. (2011). Bordering solidarities: Migrant activism and the politics of movement and camps at Calais. *Citizenship Studies*, 15(1), pp. 1–19.

Samaddar, R. (2003). *Refugees and the state: Practices of Asylum and care in India, 1947–2000*. Delhi and London: Sage.

Sanyal, R. (2012). Refugees and the city: An urban discussion. *Geography Compass*, 6(11), pp. 633–644.

Savelli, F. (2019). Roma, lite alla stazione Termini fra due senza fissa dimora. La vittima: "Io accoltellato perche avevo crocifisso". *La Repubblica*, 23 Apr.

Scholten, P. and Penninx, R. (2016). The multi-level governance of migration and integration in Europe. In: R. Penninx and B. Garces-Mascarenas, eds., *Integration of migrants into what? Integration Processes and Policies in Europe*. Dordrecht: Springer.

Siddiqui, M. (2015). *Hospitality and Islam: Welcoming in God's name*. New Haven, CT: Yale University Press.

Smith, H. (2016). *Migration crisis: Idomeni, the train stop that became 'an insult to EU values*. [online] *The Guardian*. Available at: https://www.theguardian.com/world/2016/mar/17/migration-crisis-idomeni-camp-greece-macedonia-is-an-insult-to-eu-values [Accessed 17 May 2018].

Smith, S. and Low, N. (2006). *The politics of public space*. New York: Routledge.

Spencer, S. and Delvino, N. (2019). Municipal activism on irregular migrants: The framing of inclusive approaches at the local level. *Journal of Immigrant and Refugee Studies*, 17(1), pp. 27–43.

Squire, V. and Darling, J. (2013). The 'minor" politics of rightful presence: Justice and relationality in city of sanctuary. *International Political Sociology*, 7, pp. 59–74.

Ticktin, M. (2011). *Casualties of care: Immigration and the politics of humanitarianism in France*. Berkeley, CA: University of California Press.

Viderman, T. and Knierbein, S. (2018). Reconnecting public space and housing research through affective practice. *Journal of Urban Design*, 23(6), pp. 843–858.

Villa, M. (2018). I nuovi irregolari in Italia. *ISPI Online*, 18 Dec. Available at: https://www.ispionline.it/it/pubblicazione/i-nuovi-irregolari-italia-21812

Weststeijn, A. and Whitling, F. (2017). *Termini: Cornerstone of modern Rome*. Rome: Edizioni Quasar.

Zaman, T. (2016). *Islamic traditions of refuge in the crisis of Iraq and Syria*. New York: Palgrave.

Index

Note: Page numbers in *italics* indicate figures.

Printed in the United States
by Baker & Taylor Publisher Services